CLASSICAL AND MODERN
INTEGRATION THEORIES

Probability and Mathematical Statistics

A Series of Monographs and Textbooks

Editors

Z. W. Birnbaum
University of Washington
Seattle, Washington

E. Lukacs
Catholic University
Washington, D.C.

---

1. Thomas Ferguson. Mathematical Statistics: A Decision Theoretic Approach. 1967

2. Howard Tucker. A Graduate Course in Probability. 1967

3. K. R. Parthasarathy. Probability Measures on Metric Spaces. 1967

4. P. Révész. The Laws of Large Numbers. 1968

5. H. P. McKean, Jr. Stochastic Integrals. 1969

6. B. V. Gnedenko, Yu. K. Belyayev, and A. D. Solovyev. Mathematical Methods of Reliability Theory. 1969

7. Demetrios A. Kappos. Probability Algebras and Stochastic Spaces. 1969

8. Ivan N. Pesin. Classical and Modern Integration Theories. 1970

CLASSICAL AND MODERN INTEGRATION THEORIES

IVAN N. PESIN
L'VOV UNIVERSITY
L'VOV, U.S.S.R.

Translated and Edited by SAMUEL KOTZ
DEPARTMENT OF MATHEMATICS
TEMPLE UNIVERSITY
PHILADELPHIA, PENNSYLVANIA

1970

ACADEMIC PRESS   New York and London

ACADEMIC PRESS, INC.
111 Fifth Avenue, New York, New York 10003

United Kingdom Edition published by
ACADEMIC PRESS, INC. (LONDON) LTD.
Berkeley Square House, London W1X 6BA

LIBRARY OF CONGRESS CATALOG CARD NUMBER: 71-117638
AMS 1970 Subject Classification 28-03, 28A25, 28A30

PRINTED IN THE UNITED STATES OF AMERICA

Classical and Modern Integration Theories. Translated
from the original Russian edition entitled Razvitie
Ponyatiya Integrala, published by Izdatel'stvo "Nauka"
Moscow, 1966.

To My Wife

CONTENTS

II.   THE ORIGIN OF LEBESGUE–YOUNG INTEGRATION THEORY

3. The Borel Measure

4. Lebesgue's Measure and Integration

5. Young's Integral

6. Other Definitions Related to the Definition of Lebesgue's Integral

7. Stieltjes' Integral

III. INTEGRATION IN THE SECOND DECADE OF THE 20th CENTURY

8. The Problem of the Primitive—The Denjoy–Khinchin Integral

9. Perron's Integral

10. Daniell's Integral

FOREWORD

In presenting the edited translation of I. N. Pesin's monograph "Classical and Modern Integration Theories,"[1] the following remarks seem in order.

This monograph belongs to a type of mathematical literature uncommon among conventional textbooks. It comprises a detailed historical survey of the development of classical integration theory, from Cauchy to Daniell, with a brief review of more recent models. It is in the form of *almost* verbatim quotations of basic definitions and theorems from original papers and treatises (with unified notation throughout the book), followed by commentaries and analyses of the concepts and arguments introduced, emphasizing their interrelation with previous (and in some cases future) theories and comparing the relative merits and features of various developments.

This structure results in a vivid, readable account of integration theory, and the ideas of the classical authors (Borel, Lebesgue, Young, Stieltjes, Radon, and others) are presented in the proper perspective of modern mathematics.

The only omission of some consequence is the absence of any discussion concerning an early result by V. Volterra indicating certain limitations in Riemann's integration theory [see *G. Mat.* **19**, 333–372 (1881)].

---

[1] The original title was "Development of the Concept of the Integral."

Such a book may therefore appeal to readers of various and diverse categories:

(a) It is a valuable source for those interested in the history of mathematics in the second part of the nineteenth century and the first two decades of the present century—one of its most fruitful and decisive periods.

(b) It contains useful background reading material for graduate students taking courses in the theory of functions of a real variable and related topics.

(c) The book is suitable as a textbook in integration theory for student teachers enrolled in mathematical education programs.

(d) The English translation of this book is also intended as "refresher reading" for instructors of calculus (on elementary and intermediate levels). Such a text will undoubtedly enrich their knowledge of integration theory, and contribute to sounder instruction in their courses, especially for students planning to embark on careers in the mathematical sciences.

In translating the book, we tried to follow faithfully the original Russian text with, however, several revisions and modifications. In some places, where the author's exposition may seem somewhat verbose for an American reader, we have made a few relatively short omissions. We have also corrected minor errors and misprints in the original text and made certain changes in notation. In some instances, references to Russian publications have been replaced by corresponding English texts. A few notes and bibliographical corrections have been added as well as an index of terms and authors, for the reader's convenience. Sections of the text related to side issues or technical details are starred.

This book, in the translator's opinion, is an insightful and well-motivated exposition of the classical theory of integration written by the distinguished Soviet mathematician Ivan N. Pesin, whose main contributions are in the field of quasi-conformal transformations, Riemann surfaces, and the problems related to the theory of functions of a real variable.

It is hoped that the English version will be equally stimulating and useful for anyone who wishes to understand better the problems and foundations of classical mathematical analysis.

SAMUEL KOTZ

PREFACE

I feel that it is worthwhile for a person, at the final stages of his mathematical education, to acquaint himself with the history of the development of classical mathematical ideas. This, on the one hand, is of interest on its own and, on the other, assists him in obtaining a deeper understanding of the essence of the subject. One of the main topics in the theory of functions is integration theory with its numerous interconnections. This book consists of essays on the development of the notion of the integral. In these essays I have attempted to present the development of classical integration theory, i.e., that part of the theory that is directly connected with the problems of area and the determination of primitive functions, beginning with Cauchy and concluding with the investigations conducted during the second decade of the present century. When compiling these essays I tried to avoid too general conclusions, confining myself to the most essential remarks; I tried to present the "raw material" to the reader in a manner such that he would arrive at these conclusions on his own. As a rule the outlines of the original proofs have been retained in the exposition. However, the outdated terminology and notation have been modernized. The reader should not be disconcerted if he does not find a detailed description of the works of individual mathematicians in the book—the purpose of this book is to trace the development of the notion of integration as a whole.

In view of limitations in space I was unable to present adequately the interrelation between integration and other branches of mathematics, for example, the theory of trigonometric series, a subject in which problems originated that stimulated the development of integration theory. It may seem that many of the proofs appearing in the text could have been omitted without any loss. However, it must be agreed that the theory of functions is to a large extent a certain mode of thinking and therefore its evolution is intrinsically connected with the development of its specific methods of reasoning; one cannot describe this evolution without a presentation of the arguments.

When writing this book it was necessary sometimes to deviate from a description of the arguments and results of the original authors in order to unify or simplify the exposition. As a rule, the reader can easily detect such deviations. In those cases when this may not be so obvious, the arguments that are not due to the original author are distinguished by fine print.[1] Some incidental remarks are also printed in this manner.

It is quite possible that I was not always accurate in my historical remarks, but I believe that gross errors have been avoided. The rudiments of calculus and the elements of set theory are sufficient prerequisites for the first part of this book; for the remainder, at least a superficial knowledge of measure theory and Lebesgue integrals is desirable. We assume that the reader possesses a knowledge of the theory of completely ordered sets and transfinite numbers as presented, for example, in the first five sections of Chapter 14 of I. P. Natanson's "Theory of Functions of a Real Variable," 2nd ed.[2] However, even without this it is possible, with minor exceptions, to comprehend all the sections of the first and second parts of the book.

I am thankful to all those who assisted me in this work and especially to Professor L. I. Volkovyskiĭ and Professor I. G. Sokolov—the latter initiated the idea of writing this book. G. L. Lunz not only expressed a constant and lively interest in the book, but also, as its scientific editor, contributed much to the substantive improvements of the text. I express to him my deep gratitude.

[1] Starred in this volume.

[2] Natanson, I. P., "Theory of Functions of a Real Variable," 2nd ed. Vols. 1 and 2. Ungar, New York, 1955 (Vol. 1), 1960 (Vol. 2). (English translation by L. F. Boron of the *first edition* with amendments that were included by the author in his second edition.)

NOTATION AND TERMINOLOGY

We present here the notation, terminology, and assertions which are used throughout the book.

The sum (union), product (intersection), and difference of two sets A and B are denoted $A + B$, $A \cdot B$, and $A - B$, respectively.

The null set (empty set) is denoted by 0.

R_n is the Euclidean n-dimensional space.

The symbol $E(...)$ denotes the set of points satisfying the condition appearing in the brackets.[1]

The symbol $\overset{\text{def}}{=}$ means that the left-hand side of the equality is defined by the right-hand side; for example, the equality $E \overset{\text{def}}{=} E_x(\omega(x, f) > \alpha)$ means: E is by definition the set of the point x for which the variation (see below) of the function $f(\cdot)$ is greater than α.[2]

E' is the set of all accumulation points of E.

$\bar{E} \overset{\text{def}}{=} E + E'$.

---

[1] In American literature the notation $E(...)$ is not common; rather one uses $\{\,|...\}$ or $\{\,;...\}$.

[2] An alternative, more commonly used notation is: $E = \{x | \omega(x, f) > \alpha\}$.

CE is the complement of E; if it is not clear from the context relative to which set the complement is taken, the notation $C_A E$, meaning the complement of E relative to A, is used.

$J(E)$ is the set of inner points of the set E.

A *boundary point* of the set E is a point in every neighborhood of which there are points of E and of its complement CE.

The collection of all boundary points of a set E is called the *boundary* of E; we denote it by $\Gamma(E) = \bar{E} \cdot \overline{CE}$.

If E_1 is a subset of E ($E_1 \subset E$), we say that E_1 is contained in E; if $E_1 \subset J(E)$ we say that E_1 is contained *inside* E.

The notation $G = \sum_n \delta_n$ stands for the representation of the open set G as a disjoint union of open intervals δ_n.

The notation $CF = \sum_n \delta_n$ means that $\{\delta_n\}$ is a system of adjacent intervals for the closed set F.

A *portion* $\delta \cdot E$ of a set E is the intersection of the set E with an interval δ.

An *open neighborhood* of the set E is an open set containing E.

A monotonic sequence of open sets $\{G_n\}$ *contracts* to the set E if $\prod_n G_n = E$. The set E_1 is *everywhere dense* in E if $\bar{E}_1 \supset E$.

The set E_1 is *nowhere dense* in E if every nonempty portion of E contains a nonempty portion δE such that δE_1 is empty.

Baire's category theorem: a closed set cannot be represented as a finite or countable sum of its nowhere dense subsets. (For the proof see, e.g., Natanson [1].)

Let β be a finite or transfinite number and P be a set. The *derivative* $P^{(\beta)}$ of order β is defined by means of transfinite induction as follows: if β is a finite or transfinite number of the first kind, then $P^{(\beta)} \overset{\text{def}}{=} (P^{(\beta-1)})'$; otherwise $P^{(\beta)} \overset{\text{def}}{=} \prod_{\gamma < \beta} P^{(\gamma)}$.

The set P is called *reducible* if $P^{(\beta)} = 0$, where β is a finite or transfinite number not higher than numbers of the second class; for finite β, P is said to be a set *of the first kind*. In order that a closed set P be reducible it is necessary and sufficient that it be at most countable.

An *end point* of a closed set F is the end point of an adjacent interval for the set F.

The *diameter* $d(E)$ of a set E is the supremum of the differences $x' - x''$, where $x' \in E$, $x'' \in E$.

A *partition* $\sigma = \{x_i\}_0^n$ *of a segment* $[a, b]$ is a sequence of points satisfying the inequalities $a = x_0 < \cdots < x_n = b$. The inclusion $\sigma \subset \sigma_1$ means that all the points of the partition σ are contained in the partition σ_1; the sum $\sigma_1 + \sigma_2$ is a partition consisting of the points of σ_1 and σ_2.

A partition of the set E into disjoint subsets $\{E_i\}$ is a representation of E as the sum $\sum E_i$.

The diameter of partition is $\sup_i(x_i - x_{i-1})$ [or sup $d(E_i)$]. The difference $x_i - x_{i-1}$ will be denoted by Δx_i or Δ_i; sometimes this notation will be used to denote the segment $[x_{i-1}, x_i]$.

A ring of sets is a nonempty family T of sets possessing the property that $A \in T$, $B \in T$, implies that $A + B \in T$ and $A - B \in T$.

A σ-ring of sets is a ring T of sets possessing the property that $A_i \in T$, $i = 1, 2, \ldots$, implies $\sum A_i \in T$.

An ordinate set $E(f, M)$ of a function $f \geqslant 0$ over M is the set $E_{(x,y)}\{x \in M, 0 \leqslant y < f(x)\}$.

mE denotes a measure (in some sense) of the set E; if E is a system of nonoverlapping intervals or segments, then mE is the sum of their lengths.

A measurable cover of the set E is a measurable set containing E whose measure is equal to the outer measure of E.

A point of nonintegrability (unboundedness, variability) of a function is a point in each neighborhood of which the function is nonintegrable (unbounded, nonconstant).

The oscillation $\omega(f, E)$ of the function f on the set E is the difference $\sup_E f(x) - \inf_E f(x)$.

The oscillation $\omega(f, x_0)$ of the function f at the point x_0 is $\lim_{\varepsilon \to 0} \omega(f, [x_0 - \varepsilon, x_0 + \varepsilon])$,

$$f^+ \overset{\text{def}}{=} \frac{|f| + f}{2}, \qquad f^- \overset{\text{def}}{=} \frac{|f| - f}{2}, \qquad f = f^+ - f^-.$$

f is called *absolutely integrable (on E)* if $|f|$ is integrable (on E).

$$r(f(x), x_0, x_0 + h) \overset{\text{def}}{=} \frac{f(x_0 + h) - f(x_0)}{h},$$

$$\bar{D}^+ f(x_0) \overset{\text{def}}{=} \overline{\lim_{h \to +0}}\, r(f, x_0, x_0 + h),$$

$$\underline{D}^+ f(x_0) \overset{\text{def}}{=} \varliminf_{h \to +0} r(f, x_0, x_0 + h),$$

$\bar{D}^-$, $\underline{D}^-$ are defined analogously for $h \to -0$.

$$\bar{D}f(x_0) \overset{\text{def}}{=} \overline{\lim_{h \to 0}}\, r(f, x_0, x_0 + h); \qquad \underline{D}f(x_0) \overset{\text{def}}{=} \varliminf_{h \to 0} r(f, x_0, x_0 + h).$$

If $\Delta = [a, b]$ or $\Delta = (a, b)$ and f is a continuous function, then $f(\Delta) = f(b) - f(a)$.

A function f is *continuous at the point* $x_0 \in P$ with respect to P if $f(x)$ tends to $f(x_0)$ when x tends to x_0 along P.

The characteristic function of the set E is the function which is equal to 1 on E and 0 on CE.

The Cantor–Baire stationarity principle: let a closed set F correspond to each finite or transfinite number β of the second class such that $F_\beta \supset F_\gamma$ for $\beta < \gamma$. Then there exists β_0 such that $F_{\beta_0} = F_\beta$ for all $\beta > \beta_0$. (See Natanson [1].)

The positive (negative) variation $V_a^{b+}(f)$ $(V_a^{b-}(f))$ of a function f is the supremum (infimum) of sums of nonnegative (nonpositive) summands of the form $f(x_i) - f(x_{i-1})$, where $\sigma = \{x_i\}$ is a partition of $[a, b]$ and the bounds are taken with respect to all possible partitions of the segment $[a, b]$. *The absolute variation* is the difference $V^+ - V^-$.

The Baire classification of functions: the functions of class α, where α is finite, are functions which are pointwise limits of sequences of functions of class $\alpha - 1$, but which do not belong to this class; all the continuous functions are functions of the Baire class zero. See Natanson [1] concerning the extension of this classification for the case of an arbitrary transfinite α.

The Baire theorem on functions of class one: in order that f be a function of Baire class not higher than one on $[a, b]$ it is necessary and sufficient that f have on every perfect set $P \subset [a, b]$ an everywhere dense set of continuity points with respect to P (Natanson [1]).

CLASSICAL AND MODERN
INTEGRATION THEORIES

I FROM CAUCHY TO LEBESGUE

1 FROM CAUCHY TO RIEMANN

The definition given by Cauchy is usually considered to be the first definition of an integral satisfying the modern requirements of rigor.

We shall omit an introduction which should contain a description of the period preceding Cauchy; this topic has been the subject of several investigations and has been widely discussed in the literature. We note as an example A. P. Yushkevich's [1] treatise "On the Origin of the Notion of Cauchy's Definite Integral" published in 1947.[1]

1.1 CAUCHY'S DEFINITION OF AN INTEGRAL

We present the definition of an integral given by Cauchy [1, 2]. Assuming that f is a continuous function on the segment $[a, b]$, Cauchy considers the integral sum

---

[1] Two classical and readily available references in English dealing with this subject are: BOYER, C. B., "The Concepts of the Calculus." Hafner Publ. Co., 1949. This book contains an excellent bibliography; JOURDAIN, P. E. B., The origin of Cauchy's conception of a definite integral and of the continuity of a function. *ISIS* I, 661–703 (1913). (*Translator's note.*)

$$S(\sigma) = f(x_0)(x_1 - x_0) + \cdots + f(x_{n-1})(x_n - x_{n-1}), \qquad (1.1)$$

where $\sigma = \{x_i\}_0^n$ is a partition of segment $[a, b]$. The sums S possess a "limit" as $d(\sigma) \to 0$, which is called the definite integral $\int_a^b f(x)\, dx$. Cauchy's proof of the existence of this limit is as follows: in view of the inequality

$$(x_n - x_0) \min_{[x_0,\, x_n]} f \leqslant S \leqslant (x_n - x_0) \max_{[x_0,\, x_n]} f$$

and the property of a continuous function to admit any intermediate value, one can write

$$S = f(x_0 + \theta(x_n - x_0))(x_n - x_0), \qquad 0 \leqslant \theta \leqslant 1. \qquad (1.2)$$

Equation (1.2) is valid for *any* segment $[x_0, x_n]$ on which f is continuous. We construct a new subdivision σ_1, containing σ, $\sigma_1 \supset \sigma$, and the sum $S_1 = S(\sigma_1)$. Utilizing Eq. (1.2) which is valid for each one of the segments $[x_{i-1}, x_i]$, we obtain

$$\begin{aligned}
S_1 &= (x_1 - x_0)f(x_0 + \theta_1(x_1 - x_0)) + \cdots \\
&\quad + (x_n - x_{n-1})f(x_{n-1} + \theta_n(x_n - x_{n-1})) \\
&= (x_1 - x_0)[f(x_0) + \varepsilon_0] + \cdots \\
&\quad + (x_n - x_{n-1})[f(x_{n-1}) + \varepsilon_{n-1}] \\
&= S + (x_1 - x_0)\varepsilon_0 + \cdots + (x_n - x_{n-1})\varepsilon_{n-1}.
\end{aligned}$$

The number ε_{i-1} does not exceed the oscillation of the function f on the interval $[x_{i-1}, x_i]$, hence it becomes arbitrarily small if the diameter of the partition is sufficiently small (Cauchy repeatedly uses in his arguments the property of uniform continuity of functions). Thus, S_1 differs from S by an arbitrarily small amount if the diameter of the subdivision is sufficiently small. Now if $S(\sigma_2)$ and $S(\sigma_3)$ are arbitrary sums of type (1.1), then each of them differs little from the third sum $S(\sigma_2 + \sigma_3)$ obtained using the subdivision $\sigma_2 + \sigma_3$ of the segment $[a, b]$; therefore, $S(\sigma_2)$ differs little from $S(\sigma_3)$; this completes the existence proof of the limit of sums S.

Cauchy emphasizes that the argument given above is relevant under the condition that f is continuous and bounded on a bounded segment $[a, b]$; incidentally in this case the following formula is valid:

$$\lim_{\substack{\xi_1 \to a \\ \xi_2 \to b}} \int_{\xi_1}^{\xi_2} f(x)\,dx = \int_a^b f(x)\,dx, \qquad a < \xi_1 < \xi_2 < b. \tag{1.3}$$

If one of these conditions is not satisfied, the limit of the sum (1.1) may not exist[2]; however, it is possible that these conditions are satisfied in $[\xi_1, \xi_2]$ for all $a < \xi_1 < \xi_2 < b$ and the limit (1.3) does exist. In this case (which may occur only if one of the points a, b is a point of unboundedness of f), Cauchy proposes to *define* the integral $\int_a^b f(x)\,dx$ using relation (1.3). This definition is extended in an obvious manner to the case when f is continuous on $[a, b]$ except for a finite number of points of unboundedness. An integral which exists in this sense is known as an *improper integral*. The improper integral on an unbounded interval is defined in a well-known manner.

Let c, $a < c < b$, be the only point of unboundedness of f in $[a, b]$ and suppose that the improper integral does not exist. However, it may happen that the limit

$$\lim_{\varepsilon \to 0} \left[\int_a^{c-\varepsilon} f(x)\,dx + \int_{c+\varepsilon}^b f(x)\,dx \right]$$

exists; this limit is called the *principal value* of the integral $\int_a^b f(x)\,dx$. Similarly, for integration over $(-\infty, \infty)$, the limit

$$\lim_{\varepsilon \to 0} \int_{-1/\varepsilon}^{1/\varepsilon} f(x)\,dx$$

is called the principal value of the integral $\int_{-\infty}^{\infty} f(x)\,dx$. These two cases can be combined.

The Cauchy indefinite integral $\int f(x)\,dx$ is the general solution of the differential equation

$$dy = f(x)\,dx. \tag{1.4}$$

Thus the indefinite integral is the collection of functions of the form $F(x) + \omega(x)$ where $F(x)$ is a solution of Eq. (1.4) and $\omega(x)$ satisfies the equation $\omega'(x) = 0$, i.e., ω is constant.

---

[2] Cauchy writes: "If the function $f(x)$ is not, as above, finite and continuous from $x = x_0$ to $x = X$..." (Cauchy [2], p. 131).

The integral of f over $[a, b]$ is related to the indefinite integral F by the formula

$$\int_a^b f(x)\, dx = F(b) - F(a).$$

Cauchy admits the possibility that the equation $\omega'(x) = 0$ may not be satisfied on a finite set of points; in this case ω will be a piecewise constant function given by the formula

$$\omega = \frac{c_0 + c_m}{2} + \frac{c_1 - c_0}{2} \frac{x - x_1}{[(x - x_1)^2]^{1/2}} + \cdots$$

$$+ \frac{c_m - c_{m-1}}{2} \frac{x - x_m}{[(x - x_m)^2]^{1/2}}.$$

Cauchy describes the geometric meaning of the integral $\int_a^b f(x)\, dx$ as the area bounded by the curve $y = f(x)$, the abscissa axis and the ordinates $x = a, x = b$. It follows from Cauchy's arguments that he assumes the notion of an area to be given in advance.

Summarizing our discussion about the integral defined by Cauchy, we conclude that Cauchy utilizes the process of integration only for continuous functions. When defining the integral of a discontinuous function, Cauchy restricts himself to functions which are everywhere continuous except for a finite number of points of unboundedness.

Remark. Cauchy draws attention to the fact that although in the integral sum the values of the functions are taken at the left end points of the partition segments, the proof that the limit of the sums (1.1) exists covers also the case of the general sums, when for the ith segment the value of the function is taken at some arbitrary point $\xi_i \in [x_{i-1}, x_i]$.

1.2 RIEMANN'S DEFINITION OF THE INTEGRAL (R-INTEGRAL)

Riemann's definition of the integral (Riemann [1], [2]) is the same as Cauchy's except that the value of the function is chosen in an arbitrary manner in the interval $[x_{i-1}, x_i]$. However, in variance with Cauchy's arguments, and this constitutes a basic step forward, Riemann considers the totality of all "integrable" functions (those functions to which the process of integration is applicable) and examines the necessary and sufficient conditions under which a function is integrable. Thus, ". . . in which cases does a function admit integration and in which doesn't it?" (Riemann [2], p. 237). Let ω_i be the oscillation of a function in the segment $\Delta_i = [x_{i-1}, x_i]$. Riemann asserts that in order that a function f be integrable in the segment $[a, b]$ it is necessary and sufficient that the sum

$$\sum_{i=1}^{n} \omega_i \Delta_i \tag{1.5}$$

tends to zero as the diameter of the partition $d(\sigma)$ tends to zero.[3] Let f be integrable and let Δ be the least upper bound of the sums (1.5) for all subdivisions of diameter not larger than d; let $S = \sum' \Delta_i$ be the total length of those segments Δ_i for which $\omega_i \geqslant \alpha > 0$. Then

$$\alpha S = \alpha \sum' \Delta_i \leqslant \sum_{i=1}^{n} \omega_i \Delta_i \leqslant \Delta, \qquad S \leqslant \frac{\Delta}{\alpha};$$

since $\Delta \to 0$ as $d(\sigma) \to 0$, it follows from the last inequality that *for given α the sum S of the lengths of segments in which the oscillation of the function is not less than α tends to zero.* This condition is also sufficient. Indeed, let M be the least upper bound of the function $|f|$ on $[a, b]$; then the oscillation of the function f on any segment is not greater than $2M$. The following inequalities

$$\sum' \omega_i \Delta_i < 2MS, \qquad \sum \omega_i \Delta_i - \sum' \omega_i \Delta_i < \alpha(b - a)$$

hold; here the primed sums are extended over those segments in which the variation is not less than α. Therefore,

$$\sum \omega_i \Delta_i \leqslant 2MS + \alpha(b - a). \tag{1.6}$$

Let $\varepsilon > 0$ be given. If we take $\alpha \leqslant \varepsilon/(b - a)$ first, and then a subdivision σ so refined that $S < \varepsilon/2M$, then it follows from (1.6) that the sum (1.5) will be smaller than ε; the sufficiency of the condition is thus proved.

---

[3] Several subsequent authors consider the sufficiency proof as given by Riemann to be incomplete; see, for example, Smith [1] and du Bois-Reymond [1].

As an example of a discontinuous integrable function, Riemann presents the function with an everywhere dense set of points of discontinuity

$$f(x) = \sum_{n=1}^{\infty} \frac{(nx)}{n^2} \qquad (1.7)$$

[(x) denotes here the difference between x and the nearest integer if x is not of the form $k + \frac{1}{2}$, where k is an integer; in the latter case $(x) = 0$.]

*Consider[4] this example in somewhat more detail. The function (x) has a discontinuity at every point of the form $(2k + 1)/2$, while the limit on the left is equal to $+\frac{1}{2}$, and the limit on the right is equal to $-\frac{1}{2}$; at the point itself the value of the function is zero. The function (nx) has the same properties at the points of the form $x_n^k = (2k + 1)/2n$. If $(2k + 1)/2n$ is an irreducible fraction, then the point x_n^k is a discontinuity point only for those functions (mx) for which $m = rn$, where r is an arbitrary odd positive integer. It follows from (1.7) that[5]

$$f(x_n^k + 0) = f(x_n^k) - \frac{1}{2n^2} \sum_{i=0}^{\infty} \frac{1}{(2i + 1)^2},$$

$$f(x_n^k - 0) = f(x_n^k) + \frac{1}{2n^2} \sum_{i=0}^{\infty} \frac{1}{(2i + 1)^2}, \qquad \omega(f, x_n^k) = \frac{\pi^2}{8n^2}.$$

There exists a finite number of values n such that $\pi^2/8n^2 > \alpha$; hence in every finite interval there exists a finite number of points of the form $(2k + 1)/2n$ in which the jump of the function f is larger than α, and hence for $d(\sigma)$ sufficiently small the number S can be made arbitrarily small. The function f is thus integrable.*

Riemann notes that every piecewise monotonic function possesses the same property also, namely, there exists only a finite number of points in which the jumps of the function exceed a preassigned value; therefore monotonic functions are integrable.

[4] The material set in fine print in the original Russian edition is marked off at the beginning and end by boldface asterisks in this edition. (*Translator's note.*)
[5] The term-by-term transition to the limit in (1.7) is justified in view of the uniform convergence of the series. It also follows from this fact that f is continuous at those points at which all the terms of series (1.7) are continuous.

The study of integrals by Riemann and by certain other authors was motivated by the needs of the theory of trigonometric series. As it is known, the coefficients of the trigonometric (or Fourier) series

$$\sum a_n \cos nx + b_n \sin nx$$

are expressed in terms of their sum f by the formulas

$$a_n = \frac{1}{\pi} \int_0^{2\pi} f(x) \cos nx \, dx, \qquad b_n = \frac{1}{\pi} \int_0^{2\pi} f(x) \sin nx \, dx.$$

The first rigorous proof of the possibility of expanding a function into a trigonometric series was given by (Lejeune)–Dirichlet [1] in 1829 for the class of piecewise monotonic functions. Clearly, in order to be able to write the Fourier series of the function f it is necessary first of all for the function to be integrable. In this connection Riemann, continuing Dirichlet's investigations, felt the need to analyze the notion of an integral (Riemann [1] and [2], p. 225).

1.3   UPPER AND LOWER DARBOUX INTEGRALS

It is appropriate at this point to mention Darboux's paper [1] devoted to the problems of Riemann integration, although it is more recent. Darboux proves that *for any bounded function f the sums $\sum M_i \Delta x_i$ and $\sum m_i \Delta x_i$ corresponding to it, where $M_i = \sup_{\Delta x_i} f(x)$ and $m_i = \inf_{\Delta x_i} f(x)$, tend to a limit as $d(\sigma) \to 0$*; moreover, the limit of the upper sums is equal to their greatest lower bound and the limit of the lower sums to their least upper bound. The greatest lower bound of the upper sums and the least upper bound of the lower sums are called, respectively, *the upper and the lower Darboux integrals* and are denoted by $\overline{\int_a^b} f(x) \, dx$ and $\underline{\int_a^b} f(x) \, dx$. The function f is said to be integrable if these integrals agree. This definition of integrability is equivalent to Riemann's definition.

A similar theorem was proved by Smith [1].

2 DEVELOPMENT OF INTEGRATION IDEAS IN THE SECOND HALF OF THE 19TH CENTURY

The efforts of mathematicians working in the field of integration theory in the period 1870–1890 were primarily directed toward a generalization of the process of integration to the case of unbounded functions; thus the construction of the improper integral in Cauchy's sense was extended to apply to functions which are unbounded in the neighborhood of an infinite set of points and are integrable in the intervals of their boundedness. The papers devoted to generalizations of improper integrals are closely connected with the basic notions of set theory developed in the 1870's by George Cantor. Cantor himself took part in the construction of measure theory, which was then used in integration theory. We shall first consider an earlier generalization of the improper integral which does not require the notion of a measure. The idea of this generalization is due to Dirichlet. Dirichlet himself did not write on this subject, but we learn about this generalization from

Lipschitz, who attended Dirichlet's lectures, in his reports [1] and [2] (see Lebesgue [3]–[5] in this connection).

2.1   THE IMPROPER DIRICHLET INTEGRAL (DI-INTEGRAL)

Lipschitz considered the case when the set of points of unboundedness E^∞ of the function f has a finite number of accumulation points $\{x\}_1^n$, $x_i < x_{i+1}$ (outside of E^∞ the function is assumed to be continuous). In every segment Δ not containing points x_i the function has a finite number of points of unboundedness, and therefore the integral on Δ can be defined as an improper integral in the Cauchy sense; thus integrals of the form $\int_{x_{i-1}+\varepsilon_i}^{x_i-\delta_i} f(x)\, dx$, $i = 1, 2, \ldots, n$ are well defined if $\varepsilon_i + \delta_i < x_i - x_{i-1}$. The integral $\int_a^b f(x)\, dx$ is then defined as the limit

$$\int_a^b f(x)\, dx \overset{\text{def}}{=} \lim \sum \int_{x_{i-1}+\varepsilon_i}^{x_i-\delta_i} f(x)\, dx, \tag{2.1}$$

when all ε_i, δ_i approach zero.

Thus in order for the integral in the Dirichlet sense to exist it is necessary and sufficient that two limiting processes be valid: the first is required to obtain an improper integral in Cauchy's sense on any segment $[x_{i-1} + \varepsilon_i, x_i - \delta_i]$, and the second is in the right-hand side of (2.1). The conditions for the existence of the Dirichlet integral can also be formulated in a different manner. Let α_1, α_2 be points of unboundedness of f and let f be continuous in (α_1, α_2); furthermore, let $\alpha_1 < x' < \alpha_2$, $\alpha_1 < x < \alpha_2$; then

$$F_{\alpha_1, \alpha_2}(x) \overset{\text{def}}{=} \int_{x'}^x f(x)\, dx$$

is a continuous function defined in (α_1, α_2). The existence of the first limit means that F_{α_1, α_2} is continuous in the closed interval $[\alpha_1, \alpha_2]$; taking an adjacent segment $[\alpha_2, \alpha_3]$ and constructing the function F_{α_2, α_3}, we obtain a continuous extension of the first function following the addition of a corresponding constant. This extension can be carried out on the whole sequence of adjacent segments. In the final analysis, the existence of the first limit means the following: in each interval (x_{i-1}, x_i)

there exists a continuous function F satisfying the relation

$$F(x'') - F(x') = \int_{x'}^{x''} f(x)\, dx \qquad (2.2)$$

for any segment $[x', x'']$ in which f is continuous. It follows from (2.2) that F is uniquely determined *up to an additive constant*. The condition for existence of the second limit (2.1) is the condition of continuity of F on every *closed* segment $[x_{i-1}, x_i]$. Now the definition of integrability of f in the Dirichlet sense can be formulated as follows.

Definition 2.1

The function f is integrable in $[a, b]$ if there exists a unique[1] function F continuous on $[a, b]$ satisfying condition (2.2) in every segment $[x', x'']$ where f is continuous.

In a *nonrigorous form* this type of definition for the case of a finite number of points of unboundedness appears in Dirichlet's paper (Lejeune–Dirichlet [2]) dated 1837. This definition apparently acquired its final form in Hölder's work [1]; cf. Section 2.3).

2.2 GENERALIZATION

The definition of integrability of the function f given above does not actually assume the existence of any properties for the function. Therefore, it encompasses a wider class of functions f than was assumed at the beginning (namely, that E^∞ possesses a finite number of accumulation points). This indicates a possible generalization of the Dirichlet integral. However, the requirement of *uniqueness* of the function F in Definition 2.1 leads us to the conclusion that this definition is applicable only to functions f which have no more than an irreducible set of unboundedness points (and are continuous on the complement). This fact was noted by Hölder [1] and Harnack [5]. Indeed, Cantor later constructed a continuous function θ which is constant on the intervals adjacent to a given perfect nowhere dense set, but which is not a constant. If the set E^∞ is irreducible and hence contains a perfect set, then the function

---

[1] Up to a constant; the same also applies to Definition 2.2. (*Translator's note.*)

$F + \theta$, as well as F, satisfies relation (2.2); thus there is no possibility of uniqueness as stipulated in Definition 2.1. An argument also due to Cantor shows conversely that if the set E^∞ is separable, then the function F is *unique* provided it exists. This argument is as follows: the difference between these two functions is a continuous function θ, which is constant on the intervals adjacent to E^∞, and therefore admits no more than a finite number of values on these intervals; hence this function, being continuous, must be a constant function.

2.3 FURTHER GENERALIZATIONS: HÖLDER'S INTEGRAL

A second possible direction in which the Di-integral can be generalized is based essentially on a more general interpretation of the right-hand side of (2.2). This was the approach taken by Hölder [1] who used in (2.2) Riemann's integral instead of Cauchy's integral. Here is the corresponding definition.

Definition 2.2

A function f is called integrable on the segment $[a, b]$ if the set of points T of the segment $[a, b]$ on which it is not integrable in the Riemann sense is at most countable and there exists a continuous function F satisfying relation (2.2) on every segment $[x', x'']$, where f is Riemann-integrable; the number $F(b) - F(a)$ is called the definite integral of function f.

The fact that the function F is unique if it exists follows from the fact that the set T is necessarily closed and hence, the uniqueness proof given in the previous section is applicable. We note, moreover, that the set T coincides with the set E^∞ (i.e., the point x belongs to T since f is unbounded at this point).

Hölder did not study the properties of his integral.[2]

Definitions 2.1 and 2.2 are examples of descriptive definitions of an integral, i.e., definitions in which the integral is defined by stipulating

---

[2] The generalization suggested by Hölder was not presented originally as a new definition of an integral; this generalization was required in connection with his investigations on the possibility of representing coefficients of a trigonometric series in terms of its sum by means of Fourier's formulas in the case when the latter possesses the integral F in the sense of Definition 2.2.

some of its characteristic properties without directly indicating its construction. The argument given in Section 2.1 describes the construction of the Di-integral in the case where the second derivative (E^∞) is void. This argument could have been extended by induction to the case when $(E^\infty)^{(n)}$ is void for some integer n (du Bois-Reymond [1]). In the general case, since the countable set E^∞ is reducible there exists a transfinite number ω of at most second class for which $(E^\infty)^{(\omega)}$ is void. In this case the construction of the Di-integral can be carried out using transfinite induction. Schoenflies [1] was the first to point out the possibility of a constructive definition in the general case of a reducible E. (We note that Schoenflies' book gives a survey of developments of the theory of functions of real variables and its applications in the second half of the 19th century. It is complemented by his second book [2].)

2.4 CONTINUATION

*We present this construction assuming that the function F exists. In Section 2.1 a description of the function F in the intervals adjacent to $(E^\infty)'$ is given. Assume now that in every interval, adjacent to $(E^\infty)^{(\gamma)}$ (where γ is an arbitrary finite or transfinite number smaller than ω), the function F is constructed up to an additive constant; we show how to construct F in the intervals adjacent to $(E^\infty)^{(\omega)}$. Consider, as usual, two cases. (1) The number ω is of the first kind. Then there exists a preceding transfinite number $\omega - 1$ and the set $(E^\infty)^{(\omega)}$ is the first derivative of the set $(E^\infty)^{(\omega-1)}$ in the interval adjacent to which F is known; the construction of F in the intervals adjacent to the set $(E^\infty)^{(\omega)}$ is carried out as described in Section 2.1. (2) ω is a transfinite number of the second kind. As it follows from the formula

$$(E^\infty)^{(\omega)} = \prod_{\gamma < \omega} (E^\infty)^{(\gamma)},$$

every segment Δ_1, strictly interior to the adjacent interval Δ of the set $(E^\infty)^{(\omega)}$, belongs to the complement of some $(E^\infty)^{(\gamma)}$, $\gamma < \omega$ and hence F is already defined in Δ_1 by the induction assumption. F remains to be defined in the closed interval $\bar{\Delta}$ by continuity. We have shown a method of constructing F in the interval adjacent to any $(E^\infty)^{(\omega)}$. Let ω be the transfinite number for which $(E^\infty)^{(\omega)}$ is void (ω is necessarily of the first type). Then as a result of the ωth operation we construct F on the whole segment $[a, b]$ as was required.

It is evident from the above that for the existence of the function F, it is necessary and sufficient that, in general, a transfinite sequence of operations could be carried out. For every ω of the first kind it is easy to construct an example of a function f for which the determination of F actually requires the performance of ω operations.*

2.5 SETS OF ZERO EXTENT

As was previously mentioned in Chapter 1, the new element introduced by Riemann was that he considered the totality of functions to which the integration process is applicable and formulated a necessary and sufficient criterion for integrability. However, he did not relate clearly the integrability to the degree of discontinuity of a function. (In the formulation of his criterion, the oscillation of the function at the discontinuity points does not appear.) Du Bois-Reymond [3] formulated this idea as a mathematical assertion: he showed that *if a function is such that, for each $\alpha > 0$, the set $E_\alpha \overset{\text{def}}{=} E_x(\omega(f, x) > \alpha)$ can be included in a finite system of intervals of an arbitrarily small total length, then Riemann's integrability criterion is satisfied and conversely.*[3] We shall prove this assertion.

Proof

Necessity. Every segment containing points of E_α is such that the oscillation of the function on it is not smaller than α. Therefore E_α is contained in those subdividing segments in which the oscillation is greater than α. If the function is integrable, then according to Riemann's criterion the total length of these segments tends to zero. Q.E.D.

Sufficiency. We first note that if at all the points of some segment Δ the oscillation of a function is less than α, an $\varepsilon > 0$ can be found such

---

[3] This assertion was almost simultaneously formulated by Harnack and Dini (cf. Enzyklopädie). Much earlier (in 1870) it was formulated by Hankel [1] without a satisfactory proof.

that $\omega(f, \delta) < \alpha$, where δ is an arbitrary segment contained in Δ of length smaller than ε. We enclose E into a finite system of intervals $\Delta_1, \ldots, \Delta_m$ of total length S. Denote the segments complementary to $\Delta_1, \ldots, \Delta_m$ by $\Delta_1', \ldots \Delta_m'$; these all consist of points in which the oscillation of the function is $< \alpha$. Let d be the length of the smallest of the intervals Δ_i and let ε be the smallest of the numbers ε_i chosen for the segments Δ_i' as stated above. Consider an arbitrary subdivision of the segment $[a, b]$ of diameter smaller than $\min(d, \varepsilon)$. Then those segments of a partition σ in which the oscillation of the function is $\geq \alpha$ necessarily have points in common with the intervals Δ_i. But their total length does not exceed $3S$; since the number S can be chosen arbitrarily small, the proof is thus completed.

This form of Riemann's criterion as given by du Bois-Reymond is of special interest since it shows how this condition contains the idea of a set of measure (length, extent) zero: *a bounded function is integrable if the set E_α has extent zero for any α.*

Definition 2.3

The set E is called a set of extent zero if for every $\varepsilon > 0$ there exists a finite system of intervals $\{\Delta_i\}_1^n$ of total length $< \varepsilon$ covering the set E.

Definition 2.4

The set $E \subset [a, b]$ is called a set of extent zero if the total length of those segments in the partition σ of the interval $[a, b]$ which contain points of E_1 tends to zero as the diameter of the partition tends to zero.

We leave it to the reader to show that these definitions are equivalent.

Sets of extent zero apparently appear for the first time in Hankel's paper [1]; du Bois-Reymond [3] calls them *integrable sets*, evidently desiring to stress the significance of these sets in the theory of Riemann integration. Harnack [1], [2] calls them *discrete sets*. In our exposition we shall utilize the latter terminology. (It should be noted however, that the term "discrete set" is nowadays used in a different sense).

We note several properties of discrete sets which follow directly from the definition:

(a) A subset of a discrete set is discrete (hence the difference and product of discrete sets are discrete).
(b)   A finite sum of discrete sets is discrete (Harnack).
(c)   The closure of a discrete set is discrete (Harnack).

Properties (a) and (c) show that *the class of discrete sets consists of closed discrete sets and their subsets.* Cantor [1] proved that any reducible set is discrete; however there exist perfect discrete sets (for example, the well-known ternary Cantor set). We also note that the class of closed discrete sets coincides with the class of closed sets of measure zero in the Lebesgue sense. Harnack pointed out that there are several properties of a function which do not depend on the values assumed by the function on a discrete set (for example, the property of integrability of a bounded function).

Harnack [3] cautions us against making the erroneous conclusion that the set is discrete if it can be covered by a not necessarily finite system of intervals of an arbitrarily small total length; he noted that every countable set possesses this property and tried to determine when one can conclude the existence of a finite covering, with approximately the same length, from the existence of an infinite covering.

It is important to remember that a discrete set is necessarily nowhere dense [see property (c) of discrete sets]. It is interesting that Hankel [1] assumed that the converse is also true: that every *nowhere dense* set is discrete; his error was repeated by Harnack [2]. This type of error was due to the fact that in those days the existence of *everywhere dense sets* was unknown. (The set-theoretical notions in analysis which originated in Cantor's investigations were then in their inception.) The existence of such sets was first pointed out by Smith [1] and later by du Bois-Reymond [2] and others. Du Bois-Reymond noted that a function which is zero on intervals and one on their complements is continuous on the points of the interval and hence is continuous on an everywhere dense set; however such a function is not integrable if the sum of lengths of intervals is less than the length of the segment on which these intervals are densely situated. Thus the property of the function to have an everywhere dense set of continuity points is necessary but not sufficient for its integrability in the Riemann sense.

2.6   HARNACK INTEGRALS (H-INTEGRALS)

Approximately in the middle of the last century unbounded functions started to appear in the theory of trigonometric series. It was observed that several important theorems of analysis (such as the second mean-value theorem) are valid also for "not too" unbounded functions; examples of such functions are the improperly integrable functions in the Cauchy sense or more generally in the Dirichlet sense; these functions are unbounded at the points of some reducible set. It is therefore natural to extend this type of investigation to a more complex class of functions having a discrete set of points of unboundedness. Harnack [4] in 1883 was the first to define an integral for such functions.

Definition 2.5

Let a function f be given in $[a, b]$ *with a discrete set* E^∞ *of points of unboundedness, which is integrable in the Riemann sense on every segment not containing points of* E^∞. *Let* $\{\Delta_i\}_1^n$ *be a finite system of intervals (not necessarily contained in* $[a, b]$) *containing* E^∞ *and let*

$$f_1(x) = \begin{cases} f(x), & x \bar{\in} \sum_1^n \Delta_i, \quad x \in [a, b], \\ \\ 0, & x \in \sum_1^n \Delta_i \end{cases}$$

(the function f_1 *is integrable in the Riemann sense on* $[a, b]$). *Then*

$$(\text{H}) \int_a^b f(x)\, dx \overset{\text{def}}{=} \lim_{m \, \Sigma \, \Delta_i \to 0} \int_a^b f_1(x)\, dx, \tag{2.3}$$

if the limit on the right exists. In this case the function f is called integrable in the Harnack sense (or H-integrable).*

Remark. The requirement that each interval Δ_i contains a point of E^∞ is essential; if this is not required, then every function integrable in the Harnack sense will be absolutely integrable. There is no mention of this fact in Harnack's papers, but some contemporary authors like Stolz [2] and Schoenflies [1] considered Harnack's integral to be in general conditionally convergent. The same opinion was held by the authors of Enzyklopädie.

In the case of a finite set E^∞ of unboundedness points of the function f, Harnack's integral is an improper integral in the Cauchy sense, or equivalently, a Di-integral. In the case of an infinite set E^∞, the Harnack and Dirichlet methods of integration are not equivalent as we shall see below.

Consider now the following basic properties of the integral:

(1) From the existence of the integral on $[a, b]$ follows its existence on every segment $[c, d]$, where $[c, d] \subset [a, b]$.

(2) From the existence of the integral on $[a, c]$ and $[c, b]$ follows its existence on $[a, b]$ and equality $\int_a^b = \int_a^c + \int_c^b$.

(3) If the functions f_1 and f_2 are integrable, so is $c_1 f_1 + c_2 f_2$ and

$$\int (c_1 f_1 + c_2 f_2) \, dx = c_1 \int f_1 \, dx + c_2 \int f_2 \, dx$$

(c_1, c_2 denotes constants) and

$$\int (c_1 f_1 + c_2 f_2) \, dx = c_1 \int f_1 \, dx + c_2 \int f_2 \, dx.$$

(4) $\lim\limits_{x \to a} \int_a^x f \, dx = 0.$

Harnack's integral retains the properties (1), (2), and (4) (Harnack [5], Jordan [2]). Property (3) is valid if the additional requirement of integrability of $c_1 f_1 + c_2 f_2$ is imposed. However, arithmetic operations performed over integrable functions lead in general to nonintegrable functions; in Section 2.8 (p. 26) we shall give an example of integrable f_1 and f_2, whose sum is not integrable.

*Let us prove, for example, property (1) for Harnack's integral. Let $[\alpha, \beta] \subset [a, b]$ and let the integral

$$(H) \int_a^b f(x) \, dx$$

exist. Choose two sequences of systems of segments $\{\Delta_i^k\}_1^{n_k}$ and $\{\tilde{\Delta}_i^k\}_1^{\tilde{n}_k}$, with the total length approaching zero as $k \to \infty$, which cover the discrete set of points of unboundedness of the function f on $[a, b]$ and such that the set of adjacent intervals located outside of $[\alpha, \beta]$ is the same for both systems Δ_i^k and $\tilde{\Delta}_i^k$.

Let f_1 and $\tilde{f}_1$ be functions as in Definition 2.5, constructed for the systems $\{\Delta_i{}^k\}$ and $\{\tilde{\Delta}_i{}^k\}$, respectively (k is fixed). We have

$$\int_a^b f_1(x)\,dx - \int_a^b \tilde{f}_1(x)\,dx = \int_\alpha^\beta f_1(x)\,dx - \int_\alpha^\beta \tilde{f}_1(x)\,dx. \qquad (2.4)$$

If the integral

$$(\mathrm{H}) \int_\alpha^\beta f(x)\,dx$$

does not exist, then a sequence of systems $\{\Delta_i{}^k\}$ and $\{\tilde{\Delta}_i{}^k\}$ can be chosen such that

$$\left| \int_\alpha^\beta f_1(x)\,dx - \int_\alpha^\beta \tilde{f}_1(x)\,dx \right| > \delta > 0$$

(δ is independent of n) and this in view of (2.3) and (2.4) contradicts the assumption that the function f is H-integrable on $[a, b]$. The property is thus proved.*

Stolz [2] investigated Harnack's integrals for functions integrable in their absolute value. The suspicious attitude (not without reason) which existed then toward conditionally convergent integrals can be seen from Stolz's remark in this connection ([2], p. 277). Stolz asserts that one cannot prove the basic properties of definite integrals in the case of conditional convergence; in particular Stolz believed that property (1) cannot be proved.

*As we noted earlier (Remark, p. 18), Harnack's original definition of the integral was not completely rigorous. In this connection it is interesting to note the incorrect formulation of the basic theorem given by Harnack [5], which essentially asserts that an integral is absolutely continuous: every function F which is an integral possesses the property that the limit of the sum

$$[F(y_1) - F(x_1)] + [F(y_3) - F(y_2)] + [F(y_5) - F(y_4)] + \cdots$$
$$+ [F(x_2) - F(y_n)]$$

is equal to $F(x_2) - F(x_1)$ if the points of subdivision $y_1 \ldots, y_n$ are situated inside the segment $[x_1, x_2]$ in such a manner that the sum $(y_2 - y_1) + (y_4 - y_3) + \cdots + (y_n - y_{n-1})$ tends to zero as n increases indefinitely. Also erroneous is Assertion 3 of the same paper concerning the equivalence of Definition 2.5 and Hölder's definition in the case of a reducible E^∞.*

Moore [1] in 1901 published a critical survey of definitions utilizing Harnack's (Stolz's, Jordan's) ideas. He formulated a very meaningful integrability condition: *Let $\{\delta_i\}$ be a system of adjacent intervals of the set E^∞ and let w_i be the least upper bound of the absolute values of the integrals $\int_{\delta_i} f(x)\,dx$ over all segments $\delta_i{}'$, $\delta_i{}' \subset \delta_i$. Then the convergence of the series $\sum w_i$ is necessary and sufficient for the existence of the H-integral. Moreover,*

$$\text{(H)} \int_a^b f(x)\,dx = \sum_i \int_{\delta_i} f(x)\,dx.$$

2.7 DE LA VALLÉE-POUSSIN'S INTEGRAL

In connection with a competition sponsored by the Brussels Scientific Society, de la Vallée-Poussin, in 1894, published a paper on integration. In this paper which is devoted primarily to problems of integration and differentiation under the integral sign, we find a new definition of an integral for an unbounded function.

Definition 2.6

Let f be given on $[a, b]$ and let the set E^∞ of points of unboundedness be discrete and let f be integrable in the Riemann sense in every segment not containing points of E^∞. Let N_1 and N_2 be real positive numbers and

$$f_{N_1}^{N_2}(x) = \begin{cases} f(x) & if \quad -N_1 \leqslant f(x) \leqslant N_2, \\ N_2 & if \quad f(x) \geqslant N_2, \\ -N_1 & if \quad f(x) \leqslant -N_1. \end{cases}$$

The integral in the de la Vallée-Poussin sense [$(V\text{-}P)$-integral] is defined as the limit

$$\text{(V-P)} \int_a^b f(x)\,dx = \lim_{\substack{N_1 \to \infty \\ N_2 \to \infty}} \int f_{N_1}^{N_2}(x)\,dx \tag{2.5}$$

(provided the limit exists).

Remark. From the existence of the double limit in (2.5) follows the existence of the limits in each one the the the variables N_1 and N_2 for a fixed value of the second variable.

*We now show that under the above assumptions on f the function $f_{N_1}^{N_2}(x)$ is Riemann-integrable for any finite N_1 and N_2. To this end we verify that $\omega(f_{N_2}^{N_1}, x) \leqslant \omega(f, x)$. Indeed, in every segment Δ the following inequalities hold

$$\sup_{\Delta} f_{N_1}^{N_2}(x) \leqslant \sup_{\Delta} f(x), \qquad \inf_{\Delta} f_{N_1}^{N_2}(x) \geqslant \inf_{\Delta} f(x).$$

Therefore, $\omega(f_{N_1}^{N_2}, \Delta) \leqslant \omega(f, \Delta)$. Contracting the segment Δ to the point x, we obtain the required inequality. This inequality also proves the inclusion

$$\underset{x}{E}(\omega)(f_{N_1}^{N_2}, x) \geqslant \alpha) \subset \underset{x}{E}(\omega(f, x) \geqslant \alpha). \tag{2.6}$$

Now it is sufficient to observe that the set $E_x(\omega(f, x) \geqslant \alpha) = E_\alpha$ is discrete. Indeed, the set $E_x(\omega(f, x) = +\infty)$ is discrete according to the condition on f; we enclose this set into a finite system of intervals $\{\Delta_i\}$ such that $m \sum_i \Delta_i < \varepsilon$; in the complementary segments, f is assumed to be Riemann-integrable; therefore, the part of the set E_α belonging to these segments (whose number is finite!), is discrete and can be covered by a finite system of intervals $\{\tilde{\Delta}_i\}$ of total length $< \varepsilon$. The intervals $\{\Delta_i\}$, $\{\tilde{\Delta}_i\}$ jointly cover the set E_α and their total length is $\leqslant 2\varepsilon$; therefore, E_α and hence $E_x(\omega(f_{N_1}^{N_2}, x) \geqslant \alpha)$ are discrete [cf. (2.6)] and thus $f_{N_1}^{N_2}$ satisfies the necessary and sufficient conditions of integrability. Q.E.D.*

Remark. As was noted by Schoenflies [1], the condition that the set of points E^∞ be discrete follows from the existence of the finite limit (2.5). In the case of Harnack's integral, this condition does not depend on the remaining part of the definition of the integral. (The reader should examine the consequences if the condition of discreteness of the set E^∞ in the definition of the H-integral is omitted.)

Every (V-P)-integrable function is also absolutely (V-P)-integrable. This follows from the equality $|f|_0^N = f_0^N - f_{N^0}$; the R-integrals of the function on the right-hand side exist according to the condition and hence, the function $|f|_0^N$ is R-integrable and moreover,

$$\int |f|_0^N \, dx = \int f_0^N \, dx - \int f_N^0 \, dx;$$

as $N \to \infty$, the limit on the right-hand side of the equality exists and thus $\lim_{N \to \infty} \int |f|_0^N \, dx$ also exists. De la Vallée-Poussin investigated the basic properties of his integral. First of all he emphasizes the property of absolute continuity, namely, the fact that

$$\int_{\Sigma \Delta_i} f(x) \, dx \to 0$$

as $m \sum \Delta_i \to 0$, where $\{\Delta_i\}$ is an arbitrary system of disjoint segments.

The property of absolute continuity is intrinsic to absolutely convergent integrals, its absence is characteristic of integrals which converge conditionally; for example, Harnack's integral does not possess this property.

2.8 RELATIONSHIP BETWEEN DI- AND H-INTEGRALS[4]

In this and succeeding sections we shall analyze the relationship between various types of integrals defined above.

Di- and H-integrals are conditionally convergent integrals. A Di-integral is defined for functions with a reducible set E^∞; and an H-integral for the wider class of functions with a discrete set E^∞. In the class of functions with a reducible set E^∞ to which both definitions are applicable, Di-integration is found to be more general than H-integration.

Theorem 2.1

If the function f with a reducible E^∞ is integrable in the Harnack sense, then it is integrable also in the Dirichlet sense and both integrals agree.

----

[4] This type of relationship was apparently first investigated in a fairly complete manner by Freud [1]. See also Schoenflies [1] and Enzyklopädie.

Proof

Let

$$F(x) = (\mathrm{H}) \int_a^x f(x)\, dx.$$

From the very definition of Harnack's integral, the following equality holds for every segment $[x', x'']$ in which f is R-integrable:

$$(\mathrm{H}) \int_{x'}^{x''} f(x)\, dx = (\mathrm{R}) \int_{x'}^{x''} f(x)\, dx.$$

From property (2) (see Section 2.6) of H-integrals,

$$F(x'') - F(x') = (\mathrm{H}) \int_{x'}^{x''} f(x)\, dx = (\mathrm{R}) \int_{x'}^{x''} f(x)\, dx,$$

i.e., the equality (2.2) given in Section 2.1 is satisfied. Since the function F is continuous as well [property (4) of H-integrals], the conditions of Di-integrability of function f are satisfied. The theorem is thus proved.

*We now construct an example of a function which is Di-integrable but not H-integrable. Denote by φ a function continuous in the interval $(0, 1)$ and unbounded at the points $x = 0$ and $x = 1$; let there exist, moreover, an improper integral in the Cauchy sense

$$\int_0^1 \varphi(x)\, dx = 1. \tag{A}$$

Consider furthermore the function Φ defined in the interval $(0, 1)$ and equal to

$$\frac{(-2)^{n+1}}{n+1}\, \varphi\!\left(2^{n+1}\!\left(x - \frac{1}{2^{n+1}} \right) \right)$$

for

$$x \in \left(\frac{1}{2^{n+1}}, \frac{1}{2^n} \right), \qquad n = 0, 1, 2, \ldots;$$

clearly,

$$\int_{2^{-n-1}}^{2^{-n}} \Phi(x)\, dx = \frac{(-1)^{n+1}}{n+1}.$$

The function Φ is Di-integrable on $[0, 1]$ and

$$(\text{Di}) \int_0^1 \Phi(x)\, dx = \sum_{n=0}^{\infty} \frac{(-1)^{n+1}}{n+1}. \qquad (B)$$

Consider the segments

$$\Delta_0^{(i)} = \left[\frac{1}{2^{2i+1}}, \frac{1}{2^{2i}}\right], \ldots,$$

$$\Delta_k^{(i)} = \left[\frac{1}{2^{2(i+k)+1}}, \frac{1}{2^{2(i+k)}}\right],$$

$$\Delta_{k+1}^{\prime(i)} = \left[0, \frac{1}{2^{2(i+k+1)}}\right],$$

and the complementary segments

$$\bar{\Delta}_0^{(i)} = \left[\frac{1}{2^{2i}}, 1\right],$$

$$\bar{\Delta}_1^{(i)} = \left[\frac{1}{2^{2(i+1)}}, \frac{1}{2^{2i+1}}\right], \ldots,$$

$$\bar{\Delta}_{k+1}^{(i)} = \left[\frac{1}{2^{2(i+k+1)}}, \frac{1}{2^{2(i+k)+1}}\right].$$

We have

$$\int_{\sum_{j=0}^{k+1} \bar{\Delta}_j^{(i)}} \Phi(x)\, dx = \sum_{n=0}^{2i-1} (-1)^n \frac{1}{n+1} + \sum_{n=i}^{i+k} \frac{1}{2(n+1)}. \qquad (C)$$

Clearly, the quantity $m(\sum_1^{k+1} \Delta_j^{(i)})$ can be made arbitrarily small if we choose i sufficiently large; nevertheless the integral

$$\int_{\sum_1^{k+1} \Delta_j^{(i)}} \Phi(x)\, dx$$

becomes arbitrarily large for sufficiently large k in view of the second term in the right-hand side of (C). Therefore, Φ is not integrable in the Harnack sense.

We have noted in Section 2.6 that f_1 and f_2 can be H-integrable, without the sum $f_1 + f_2$ being H-integrable. We now construct a corresponding example. Put $f_1 = \Phi_1$, where Φ_1 is constructed using some function φ_1 in the same manner as the function Φ was constructed using φ in the previous example.

We shall assume that function φ_1 is bounded on $[0, 1]$ and satisfies equality (A). Then the Harnack integral of the function f_1, which has only one point of unboundedness $x = 0$, exists and is equal to the right-hand side of (B). In place of f_2 we take a function which coincides in

$$\left[\frac{1}{2^{n+1}}, \frac{1}{2^n} \right], \qquad n = 0, 1, 2, \ldots$$

with the function

$$\frac{(-2)^{n+1}}{(n+1)^2} \, \varphi\left(2^{n+1}\left(x - \frac{1}{2^{n+1}} \right) \right),$$

where φ is the function defined in the preceding example; the function f_2 is not bounded at the points of the form $1/2^n$, however, it has on the segment $[0, 1]$ an absolutely convergent H-integral (equal to $\sum_{n=0}^{\infty} (-1)^{n+1} 1/(n+1)^2$). The sum $f_1 + f_2$ has a conditionally convergent Di-integral and is not integrable in Harnack's sense. In order to verify this it is sufficient to repeat the argument which proves the nonintegrability in Harnack's sense of the function Φ given in the previous example.*

2.9   RELATIONSHIPS BETWEEN H- AND (V-P)-INTEGRALS

If a function is conditionally integrable in the Harnack sense, then it is not integrable in the de la Vallée-Poussin sense. (This is a corollary of Theorem 2.2, which is proved below.) In the case of absolute integrability, however, both methods of integration are equivalent.

Theorem 2.2

If a function f is absolutely integrable in the Harnack sense, it is also absolutely integrable in the de la Vallée-Poussin sense and conversely.

The proof of this theorem is subdivided into several assertions.

Assertion 1

Let functions φ and f have on a given segment $[a, b]$ discrete sets E^∞, and let them be R-integrable in the interior of the adjacent intervals. If $|\varphi| \leqslant |f|$ and $|f|$ is integrable in the Harnack or de la Vallée-Poussin sense then $|\varphi|$ is also integrable in the same sense.

The proof follows directly from the definitions of the corresponding integrals. In particular if $|f|$ is integrable then f^+ and f^- are integrable and conversely; it follows from here that it is sufficient to prove Theorem 2.2 for nonnegative functions only.

Assertion 2

In order for Harnack's integral of a nonnegative function f to exist, it is sufficient that the limit (2.3) exist for some monotonic sequence $T_n = \sum_i \Delta_i^n$ of systems of intervals such that

$$T_n \supset E^\infty \qquad \text{and} \qquad \lim_{n \to \infty} T_n = E^\infty.$$

Proof

Let

$$I = \lim_{n \to \infty} (\text{R}) \int_{CT_n} f(x)\, dx$$

(CT_n is the complement of T_n relative to the segment of integration) and let $\tilde{T}$ be an arbitrary finite system of intervals containing E^∞. Let n be the largest integer for which the inclusion

$$T_n \supset \tilde{T}$$

holds. Clearly, $n \to \infty$ if $m\tilde{T} \to 0$; therefore, from the inequality

$$\int_{CT_n} f \leqslant \int_{C\tilde{T}} f \leqslant I,$$

the validity of the assertion follows.

Assertion 3

If $f \geqslant 0$ and the function f is integrable in the Harnack sense, then it is integrable in the de la Vallée-Poussin sense.

Proof

We have seen in Section 2.7 that if f is R-integrable on segments which are disjoint of some discrete set, then $f_0{}^N$ is R-integrable for every N. From the inequality

$$(\text{H}) \int_a^b f(x)\, dx \geqslant (\text{R}) \int_a^b f_0{}^N(x)\, dx,$$

the existence of a finite limit in the right-hand side of this inequality as $N \to \infty$ follows.

Assertion 4

If $f \geqslant 0$ and is integrable in the de la Vallée-Poussin sense, then it is integrable in the Harnack sense.

Proof

Let $\{T_k\}$ be a monotonic sequence of a system of intervals such that $E^\infty = \prod_1^\infty T_k$. Then, there exist numbers N_k such that $f(x) < N_k$, provided $x \in CT_k$ and hence, $f(x) = f_0^{N_k}(x)$, $x \in CT_k$. From here

$$(\text{R}) \int_{CT_k} f(x)\, dx \leqslant (\text{R}) \int_a^b f_0^{N_k}(x)\, dx \leqslant (\text{V-P}) \int_a^b f(x)\, dx.$$

From this inequality we obtain the existence of the limit, as $k \to \infty$, of the integral on the left-hand side which is equal in view of Assertion 2 to the Harnack integral of the function f.

2.10   CONDITIONALLY CONVERGENT (V-P)-INTEGRALS

De la Vallée-Poussin [1] also considers conditionally convergent integrals. The construction of an integral in this case is carried out using Dirichlet's method, and it is assumed that the *points of conditional (V-P)-integrability* form a reducible set of the first kind; in each segment, disjoint of this set, the function is assumed to be absolutely integrable in the de la Vallée-Poussin sense. Clearly, if one omits the nonessential requirement that the reducible set of points of conditional integrability be a set of the first kind, then the integral under consideration will be more general than the Dirichlet–Hölder integral.

2.11   MEASURE OF SETS—PEANO–JORDAN MEASURE

The discovery of sets of zero extent naturally led to the notion of "measure" of a set. This measure was defined using the already available method of covering sets by certain elementary regions whose length (area, volume) are assumed to be known and then passing to the limit. The properties of sets of positive measure were usually contrasted with those of discrete sets; but while the latter originated as a result of the requirements of the theory, the significance of sets of positive measure was not as yet clearly understood. The understanding came later and was due to the successes achieved by the theory of functions of a real variable in the beginning of this century.

The first definition of the measure (*Inhalt*) of an arbitrary set was given by Cantor (in 1883) and Stolz (in 1884). Somewhat later, an equivalent definition was suggested by Harnack (in 1885). These definitions were substantially supplemented by Peano (in 1887) and Jordan (in 1892). This measure is therefore usually called the Peano–Jordan measure. We present here the definitions of Cantor, Stolz and Harnack, and concentrate on the Peano–Jordan notion of measurability.

Cantors' Definition [2], [3]

Let M be an arbitrary bounded set in the space R_n. Let $S(\rho, x)$ denote the closed sphere centered at x with radius $\rho > 0$.[5] Consider finite disjoint unions $\prod(\rho)$ of spheres $S(\rho, x)$ with ρ fixed and x in the closure $\overline{M}$ of M, and let $F(\rho)$ be the total volume of $\prod(\rho)$; $F(\rho)$ decreases monotonically with ρ, and the number

$$J(M) \overset{\text{def}}{=} \lim_{\rho \to 0} F(\rho)$$

is called the measure of the set M.

Remark. The structure of the parts constituting the set $\prod(\rho)$ is not simple enough to consider the volume of these parts as elementary as was assumed by Cantor: "... Jedes dieser Stücke ein n-dimensionales

---

[5] In the case of space R_1, the sphere is replaced by a segment with center at x.

Continuum mit dazu gehöriger Begrenzung darstellt...." In order to simplify Cantor's definition, Schoenflies [2] proposes to use open spheres $S(\rho, x)$. By Borel's lemma, one then selects a finite system of open spheres covering M; in this case the volume of the configuration consisting of a finite system of spheres is determined in an obvious manner.[6]

Cantor notes the following properties of a measure:

(1) $J(M) = J(M^{(\gamma)})$, where γ is either a positive integer or a transfinite number.

(2) $J(M_1 + M_2) = J(M_1) + J(M_2)$ if the distance between the sets M_1 and M_2 is positive.

*The nonrigorous formulation of property (2) as given by Cantor is essentially an intuitive description of the conditions for which the desired equality holds: he requires that M_1 and M_2 be situated in "completely separated" (*völlig getrennt*) n-dimensional parts of the space. Cantor notes that if these conditions are violated the equality in general does not hold.*

Cantor makes a very interesting remark (from the point of view of the subsequent development of measure theory) on the possibility of defining a generalized measure, by assuming

$$F(\rho) = \int_{\Pi(\rho)} f(x_1, \ldots, x_n) \, dx_1 \cdots dx_n$$

(where f is a positive function). This idea, having subsequently developed, led after several decades to very general constructions.

Stolz's Definition [1]

Let E be a set contained in $[a, b]$. Let $\{\sigma_n\}$ be a sequence of partitions with diameter approaching zero and $\sigma_n \subset \sigma_{n+1}$. Furthermore, let $\Delta_i(\sigma_n)$ be the subdividing segments containing the points of E. The limit $\lim T(\sigma_n) = T$, where $T(\sigma_n) = \sum_i m\Delta_i(\sigma_n)$ is called the measure of the set E.

[6] The properties of the function $F(\rho)$ were investigated very recently! [See H. Fast, *Fund. Math.* **46**, (1958).]

Stolz proves the independence of the number T of the sequence $\{\sigma_n\}$ and proves furthermore that the number T is the limit of the numbers $T(\sigma_n)$ as the diameter of the subdivision α tends to zero. We present the corresponding proof:

Let the number T be obtained as above by means of some (not necessarily monotonic) sequence of partitions $\{\sigma_n\}$. Let σ' be an arbitrary partition of $[a, b]$ consisting of n' segments. If $d(\sigma_n) = \varepsilon_n$, then the total length of the subdividing segments in σ_n containing in their interior the endpoints of segments of the partition σ' does not exceed $n'\varepsilon_n$; since every segment $\Delta_i(\sigma_n)$ has common points with some segment $\Delta_i(\sigma')$, $T(\sigma_n) \leqslant T(\sigma') + \varepsilon_n n'$; letting $n \to \infty$, we obtain

$$T \leqslant T(\sigma'). \tag{2.7}$$

If T' is the limit of the sequence of numbers $T(\sigma_n')$ for some sequence of partitions $\{\sigma_n'\}$ with diameter approaching zero, we obtain from (2.7)

$$T \leqslant T'. \tag{2.8}$$

Since T and T' can be interchanged in (2.8) it follows that $T = T'$.

Q.E.D.

Stolz also extended his definiton of measure to the case of planar sets.

Harnack's Definition [3]

Let $E \subset [a, b]$, $b - a = l$. Harnack proposed the following process for defining a measure: *delete from $[a, b]$ the nonoverlapping intervals of length $\geqslant l/2$ that do not contain points of E (provided there are such intervals); from the remaining segments of total length l_1 delete the nonoverlapping intervals of length $l/3$ that do not contain points of E (provided there are such intervals); and, in general, in the nth step delete from the finite system of segments of total length l_{n-1} containing the set E, all the nonoverlapping intervals of length $\geqslant l/(n + 1)$ that do not contain points of E. The limit of the sequence of numbers l_n is the measure of the set E.*

This process of defining the measure of a set, which is due to Harnack, is uniquely defined at each step if E is a set on the real line, in the sense that there exists a unique system of nonoverlapping intervals, which do not contain points of the given set whose length exceeds a given positive number; such a system is also finite. One can similarly define a measure for sets in the plane and in spaces of higher dimension; however, in doing so, the above-mentioned uniqueness is lost. Harnack notes this fact and to avoid it he recommends that Stolz's definition be used for higher dimensional spaces.

The Peano–Jordan Definition

The Peano–Jordan definition applies to any dimension; we present the case of the plane as in Jordan [1]. Consider a square grid in the plane, consisting of straight lines parallel to the coordinate axis.

Let S be the sum of the areas of those (closed) squares of the grid which are located inside the set E [i.e., in J(E)]; let S' be the sum of areas of those squares which contain at least one boundary point of the set E. The sum S + S' is the total area of the squares containing points of the closure E + E'. As the diameter of the grid tends to 0, the numbers S and S + S' tend to limits; the first of the limits is called the inner measure of the set E, and the second is called the outer measure of the set E. When these numbers agree, the set is called measurable in the Peano–Jordan sense [or P-J-measurable], and the common value of the inner and outer measures is called the Peano–Jordan measure of the set E.

This definition is due to Jordan [1]; Peano [1] defined the outer and inner measures of a set as the infimum of the areas of a finite system of polygons containing the set and, correspondingly, the supremum of the areas of polygons contained in the set. The equivalence of these two definitions can be established in an elementary manner. In what follows, we shall call these measures Peano–Jordan measures.

Several interesting ideas can be found in Peano [1]. The fact that these ideas are not always supplemented by proofs cannot be sufficient reason for not associating them with Peano's name. Some of the highlights are the introduction of two measures of a set (singling out the case of measurability), a necessary and sufficient criterion for measurability, an interpretation of the Riemann integral as the measure of a certain ordinate set (see Section 2.13), the introduction of finitely additive set functions (*funzione distributiva*), and the notion of a derivative of one set function with respect to another. It is true, however, that Peano did not sufficiently emphasize the importance of the additive properties of a class of sets, although this additivity appears in his arguments; moreover his definition of a derivative is so rigid that (as proved by Peano) every derivative turns out to be continuous. Finally we point to Peano's definition of the integral of a real-valued function l with respect to an arbitrary nonnegative set function $\chi(E)$, as the common value of two quantities: the infimum of the upper sums $\sum_i M_i \chi(E_i)$ and the supremum of the lower sums $\sum m_i \chi(E_i)$, where $E = \sum E_i$ is a partition of the domain of integration E. Note also Peano's proof of the additivity of the integral and its differentiability as a set function in the case of a continuous l.

We do not intend to overestimate the importance of Peano's results in the theory of additive set functions and integration; as we shall see in the following text, one of the fundamental points in this theory is the proof of the existence of a sufficiently general additive set function and of the corresponding additive class of sets. However, Peano's ideas are important as ideas per se; in his time the implementation of these ideas was not on the agenda of mathematical activity. (The same was the case, when Cauchy gave a general definition of an integral, but used this definition only for continuous functions; there was no practical necessity at that time to integrate discontinuous functions.)

2.12 PROPERTIES OF THE PEANO–JORDAN MEASURE

We have seen that the measures defined by Cantor, Stolz, and Harnack agree with the outer PJ-measure.[7] The introduction of the inner measure was a new and very important feature of the Peano–Jordan definition. This notion is not however entirely new: for the case of an ordinate set the inner measure is the lower Darboux integral and is a natural generalization of it for the case of arbitrary sets; moreover, the

[7] One should note that neither Stolz nor Harnack indicate in their definition whether intervals (open) or segments (closed) are used; the term "interval" was used in both senses. The reader can verify easily that for these definitions it makes no difference whether intervals or segments are considered.

inner and outer measures of a set E are precisely the lower and upper Darboux integrals of the characteristic functions of the set! The measurability of a set means the R-integrability of the characteristic function and conversely. The PJ-measure possesses on the class of PJ-measurable sets several properties of length, and is a natural generalization of this concept. We shall discuss the properties of PJ-measures in some detail.

In what follows, the numbers $m_i E$ and $m_e E$ will denote respectively, the inner and outer Jordan measures. The following properties of these numbers were noted by Jordan: (a) $E_i \subset E_2$ implies that $m_i E_1 \leqslant m_i E_2$, $m_e E_1 \leqslant m_e E_2$; (b) $m_e(\sum_1^n E_j) \leqslant \sum_1^n m_e E_j$; (c) $m_i(\sum_1^n E_j) \geqslant \sum_1^n m_i E_j$, provided the E_j are pairwise disjoint. Other properties of the measure were pointed out by later authors. (A detailed exposition of the theory of Jordan's measure is given by Pierpont [1]; see also Hausdorff [1] and Lebesgue [3]–[5].)

*We state now several assertions concerning the properties of Jordan's measure. Assume that the set E is included in the square T, whose sides belong to the given grid and let $CE = T - E$.

Assertion 1

A necessary and sufficient criterion for measurability: *In order that the set E be PJ-measurable it is necessary and sufficient that the set of its boundary points be discrete* (i.e., of PJ-measure zero).

Proof

Since $m_e E \stackrel{\text{def}}{=} \lim(S + S')$, and $m_i E \stackrel{\text{def}}{=} \lim S$, in order that $m_i E = m_e E$, it is necessary and sufficient that $\lim S' = 0$, i.e., the set $\Gamma(E)$ be discrete.

Using this criterion for measurability it is easy to obtain various examples of measurable sets; in particular, finite systems of polygons are measurable. An important example of measurable sets of PJ-measure zero are closed sets of Lebesgue measure zero, i.e., such sets which can be included into a system of squares (not necessarily finite) with an arbitrarily small total area.

Assertion 2

$$m_i E = m_i J(E), \ m_e E = m_e \bar{E}.$$

The proof follows directly from the definitions. In particular, $mE = mJ(E) = m\bar{E}$, if E is measurable.

Assertion 3

For every $E \subset T$ the equality $m_i E + m_e CE = mT$ is valid.

Proof

The squares which form the subdivision σ, are subdivided into two groups: the group T_1 of squares contained inside E and the group T_2 of squares intersecting $\overline{CE}$. Since $mT_1 + mT_2 = mT$, the assertion follows.

Corollary

If the set E is measurable, so is CE and conversely.

Assertion 4

The sum of two measurable sets is measurable.

The proof follows from the inclusion $\Gamma(E_1 + E_2) \subset \Gamma(E_1) + \Gamma(E_2)$ and Assertion 1.

Assertion 5

PJ-measurable sets form a ring.

The proof follows from Assertions 3 and 4.

Assertion 6

If $E = E_1 + E_2$, then $m_e E \leqslant m_e E_1 + m_e E_2$ and if moreover, $E_1 \cdot E_2 = 0$ then $m_i E \geqslant m_i E_1 + m_i E_2$. The proof of the first inequality follows from the relation $\bar{E} = \bar{E}_1 + \bar{E}_2$. The proof of the second from the relation $J(E) \supset J(E)_1 + J(E_2)$.

Corollary

If E_1 and E_2 are PJ-measurable and $E_1 \cdot E_2 = 0$, then $mE = mE_1 + mE_2$. In other words: *the Jordan measure is finitely additive on the ring of PJ-measurable sets.*

Assertion 7

If E_1 is PJ-measurable and E_2 is congruent[8] to E_1, then E_2 is PJ-measurable and $m(E_2) = m(E_1)$.

Assertions 6 and 7 indicate the substantial advantage of the Peano–Jordan measure: it is defined on a ring of sets on which it is additive and congruent, i.e., it possesses the essential characteristics of length.

The simplest examples of sets not measurable in the Peano–Jordan sense are the everywhere dense sets without inner points, e.g., the set T_1 of points of a square T with rational coordinates and the set $T_2 \overset{\text{def}}{=} T - T_1$; we have here $m_e T_1 = m_e T_2 = 1$, $m_i T_1 + m_i T_2 = 0$, hence, $m_e T_1 + m_e T_2 > mT$, $m_i T_1 + m_i T_2 < mT$. *Thus the outer and inner PJ-measures are in general nonadditive.**

In the beginning of his treatise [1] "Remarques sur les intégrales définies," Jordan talks about the fact that as a result of a number of previous investigations the conditions under which a function is integrable in the Riemann sense are now completely determined. However, the significance, in Riemann's procedure, of the fact that the domain of integration is the segment $[a, b]$ is unclear. Jordan notes that two properties are substantially utilized in the construction of an integral: the first is that the (planar) domain of integration E has an area and the second is that this area possesses the property of (finite) additivity—if $e_1, \ldots, e_n$ is a partition of E into nonoverlapping subsets, then the area of E is equal to the sum of areas of the sets e_i. In an effort to preserve only these essential characteristics of the domain of integration, Jordan arrives at the notions of measurability and measure.

*We have seen above how Jordan solved this problem; it is worthwhile to note that he emphasizes nowhere the fact that the measure must possess the property of finite additivity. Apparently at that time he did

[8] Namely, E_1 can be superposed on E_2 by Euclidean translation and rotation.

not pay much attention to the profound difference between the properties of finite and countable additivity; therefore, in his treatise and in the "Cours d'Analyse" ([2], Vol. I), the additivity property of the measure is formulated in a manner which does not indicate what type of additivity he has in mind.*

We now present the general construction of the Riemann integral due to Jordan.

Let $f(x, y)$ be a bounded function on a PJ-measurable set E, $\inf_E f(x, y) = m$, $\sup_E f(x, y) = M$. Let $\prod = \{e_i; i = 1, \ldots, n\}$ be a partition of E, i.e., $E = \sum_{i=1}^{n} e_i$, where e_i are disjoint PJ-measurable sets. Let

$$M_i = \sup_{e_i} f(x, y), \qquad m_i = \inf_{e_i} f(x, y).$$

and define $d(\prod) = \sup_{1 \leqslant i \leqslant n} d(e_i)$. We construct the generalized Darboux sums: the upper sum $\bar{S}_\Pi = \sum M_i me_i$ and the lower sum $S_{-\Pi} \leqslant = \sum m_i me_i$.

Assertion 8

The limits of the upper and lower sums exist as $d(\prod) \to 0$.[9]

Proof

Let $T = \inf \bar{S}$ and let the sum $\bar{S} = \sum_1^n M_i me_i$ be such that $\bar{S} - T < \varepsilon$. Let $\bar{S}' = \sum_1^{n'} M_i' me_i'$ be an upper sum, and let $e_j' = e_{ij}$, if $e_j' \subset e_i$; if e_j' is not contained in any one of e_i, we denote it by $\tilde{e}_j'$. Let $M_{ij} = \sup_{e_{ij}} f(x, y)$, then

$$\bar{S}' = \sum_{i,j} M_{ij} me_{ij} + \sum_i \tilde{M}_i' m\tilde{e}_i'$$

$$\leqslant \sum_i \left(M_i \sum_j me_{ij} \right) + \sum_i \tilde{M}_i' m\tilde{e}_i'$$

$$\leqslant \bar{S} - \sum_i M_i \left(me_i - \sum_j me_{ij} \right) + M \sum_i m\tilde{e}_i'.$$

Taking into account that $\sum_i m\tilde{e}_i = \sum_i (me_i' - \sum_j me_{ij})$, we obtain

$$\bar{S}' \leqslant S + [M - m] \sum_i \left(me_i - \sum_j me_{ij} \right)$$

$$\leqslant T + \varepsilon + [M - m] \sum_i \left(me_i - \sum_j me_{ij} \right).$$

---

[9] In the original edition, $d(e_i)$ is used instead of $d(\prod)$. (*Translator's remark.*)

If the diameters of the sets e_i' are sufficiently small, then the sum $\sum_i (me_i - \sum_j me_{ij})$ is less than ε^{10} and

$$T \leqslant \bar{S}' \leqslant T + \varepsilon(M - m + 1),$$

which proves the convergence of the sums $\bar{S}'$ to T. The convergence of the lower sums to their upper bound can be proved analogously. The assertion is thus proved.

2.13 RIEMANN INTEGRAL—GEOMETRICAL DEFINITION

Let $E(f, [a, b])$ be the ordinate set of the function f over $[a, b]$ (i.e., the planar set of points whose coordinates are all possible values $x \in [a, b]$ and $0 \leqslant y < f(x)$). The following theorem is valid (Peano [1], Hausdorf [1], Lebesgue [3]–[5]): *In order that a bounded nonnegative function f on $[a, b]$ be R-integrable, it is necessary and sufficient that the set $E(f, [a, b])$ be PJ measurable; moreover*

$$\int_a^b f(x)\, dx = mE(f, [a, b]). \tag{2.9}$$

Proof

Necessity. Let f be integrable; then the lower Darboux sum $\sum m_i \Delta x_i$ represents the area or the Peano–Jordan measure of some polygon $T_* \subset E(f, [a, b])$; in the same manner the upper Darboux sum is the PJ-measure of a polygon $T^* \supset E(f, [a, b])$. Since the function is integrable, the difference $mT^* - mT_*$ is arbitrarily close to zero provided

---

[10] This statement requires a proof. It is sufficient to show that every summand of the sum tends to zero (n is constant). Since $m\Gamma(E) = m\Gamma(e_i) = 0$, putting $E = G_1$, $e_i = G_2$, we assume that G_1 and G_2 are open sets and $G_2 \subset G_1$ and $e_1', \ldots, e_m'$ is a subdivision of G_1 into PJ-measurable sets: $G_1 = \sum_1^m e_i'$; we prove then that $mG_2 - m\sum_j e_{ij} \to 0$ as the diameter of the subdivision tends to zero, where e_{ij} are those e_i' which fall into G_2. Indeed, it is always true that $m\sum e_{ij} \leqslant mG_2$. We choose a sufficiently refined grid σ such that $mG_2 - \sum_1^k m\sigma_i < \varepsilon$, where σ_i ($i = 1, \ldots, k$) are the squares of the grid for which $\sigma_i \subset G_2$. Let the distance $\rho(\sum_1^k \sigma_i, CG_2) = \alpha > 0$. Then, if $d(e_{ij}) < \alpha$, $m\sum e_{ij} \geqslant \sum_1^k m\sigma_i mG_2 - \varepsilon$; thus, finally $mG_2 \geqslant m\sum e_{ij} \geqslant mG_2 - \varepsilon$ for $d(e_{ij}) < \alpha$, which proves the convergence of $m\sum_j e_{ij}$ to mG_2.

the partition of the segment $[a, b]$ is sufficiently refined: since $\overline{T^* - T_*}$ $\supset E_{x,y}[y = f(x)]$, we conclude that $m\Gamma(E(f, [a, b]) = 0$ and hence, in view of Assertion 1 of Section 2.12, the ordinate set is measurable and, evidently, the equality (2.9) is satisfied.

Sufficiency. Let $E(f, [a, b])$ be measurable. Consider an arbitrary partition σ of the segment $[a, b]$; the polygons T^* and T_* constructed for this partition can be considered as sums of rectangles of a certain grid; the same is applicable to the difference $T^* - T_*$. In order to be able to assume that the diameter of this grid is arbitrarily small, as the diameter of the partition tends to zero, we perform additional horizontal cross sections. Thus, $\sum_i M_i \Delta x_i - \sum_i m_i \Delta x_i \to 0$ and the upper and lower Darboux integrals of the function f agree, which proves the R-integrability of the function.

In the general case of a bounded function of an arbitrary sign we have

$$\int_a^b f(x)\, dx = mE(f^+, [a, b]) - mE(f^-, [a, b]).$$

2.14   PIERPONT'S DEFINITION

Subsequently, Pierpont [1] studied in great detail properties of integrals of bounded functions. He gave two definitions which are modifications of Jordan's definition. The first is equivalent to Jordan's, the second is more general. We now present these definitions.

Definition 2.7   (Pierpont [1], p. 356)

Let $f(x, y)$ be bounded on a PJ-measurable set E, belonging to a square T. Let σ be a partition of T by means of a rectangular grid. Let $\{\delta_i\}$ be the set of rectangles of the grid all of whose points lie in E and let $(\xi_i, \eta_i) \in \delta_i$. The limit of the sums $\sum f(\xi_i, \eta_i) m\delta_i$ is called the integral of the function $f(x, y)$ with respect to the set E.

The reader can easily verify the equivalence between this and Jordan's definition.

Definition 2.8

Let $f(x, y)$ be a bounded function on the set E, $E \subset T(T$ is a square) and let $\{\delta_i\}$ be the rectangles of the grid containing at least one point of E; let $m_i = \inf_{\delta_i E} f(x, y)$, $M_i = \sup_{\delta_i E} f(x, y)$. Then the sums $\sum m_i m\delta_i$, $\sum M_i m\delta_i$ converge as the diameter of the grid tends to zero; when these limits agree, $f(x, y)$ is called integrable and their common value is called the integral of the function $f(x, y)$ over the set E.

The greater generality of this definition as compared with Jordan's definition is due to the fact that the measurability of the domain of integration is not assumed. Thus, Pierpont developed Jordan's idea (see p. 37) and noted quite correctly that in the definition of the integral over a given set E, it is not necessary to consider partitions of the domain of integration into parts for which the measure is additive or even defined *independently of* E. Such subdivisions do not exist if E is nonmeasurable, however *relatively measurable* partitions on which the outer PJ-measure is additive (for example, those considered in Definition 2.8) always exist.

*2.15   INDEFINITE INTEGRALS[11] AND PRIMITIVE FUNCTIONS

The classical definition of a primitive function is as follows: a primitive function F for the given function f is a function satisfying the equation

$$F'(x) = f(x) \tag{2.10}$$

for all points x (see Section 1.1).

When considering possible generalizations of the notion of a primitive function, the following two of its basic properties should be kept in mind:

(1)   The primitive function is unique up to an additive constant.
(2)   Every derivative function possesses a primitive.

[11] Here and in what follows, an indefinite integral (in any sense) is regarded as a definite integral with a variable upper limit.

For every continuous function f, its indefinite integral in the Cauchy sense $\int_a^x f(x)\,dx$ is a primitive, and the definite Cauchy integral of a continuous function f on the segment $[a, b]$ can be defined as an increment of the primitive F:

$$\int_a^b f(x)\,dx = F(b) - F(a).$$

The actual computation of an indefinite integral is based on this observation. However, this relation between primitives and definite integrals is lost as soon as we leave the realm of continuous functions (and hence, integration in the Cauchy sense).*

II THE ORIGIN OF LEBESGUE–YOUNG INTEGRATION THEORY

The discoveries in analysis in the early part of the 20th century characterize this era as the beginning of the modern theory of functions. In subsequent years these new methods penetrated deeply into various branches of mathematics.

Investigations in integration theory in the second part of the 19th century were sporadic and often motivated by diverse problems in analysis, and can be considered the spadework which led to a significant breakthrough in this area.

We have seen how the concept of integration is connected with the concept of measure. The success of the new theory was achieved by complete withdrawal from classical definitions; the decisive factor was the discovery of a class of sets far more general than those considered previously. This new class of sets is associated with a *countably* additive measure defined on it. It was this transition from finite additivity to countable additivity that ensured the success of the new theory.

3 THE BOREL MEASURE

In 1898, Emile Borel's book [1] entitled "Leçons sur la Théorie des Fonctions" appeared. From now on, we shall refer to this work as *Leçons*. On pp. 46–48, Borel formulated postulates which became the guidelines for defining measures of sets. They are as follows:

(1) *A measure is always nonnegative.*

(2) *The measure of a sum of a finite number (of nonoverlapping) sets equals the sum of their measures.*

(3) *The measure of the difference of two sets (a set and a subset) is equal to the difference of their measures.*

(4) *Every set whose measure is not zero is uncountable.*

[It follows from Condition (2) that Condition (4) is equivalent to the condition: the measure of a set consisting of one point is zero.]

Obviously, these postulates describe essential properties of length. A definition of an object in terms of the properties it should possess for certain purposes is usually called a *descriptive* definition, to be distinguished from a *constructive* definition which contains the construction of the defined object. Thus the whole of Postulates (1)–(4) should be considered a descriptive definition of the measure. However, in this connection (as in the case of any other descriptive definition), the question of existence arises: do the defined objects exist? Is there a (nontrivial) class of sets for which it is possible to define a measure which agrees with the postulates stated above? And furthermore, if a measure exists, is it unique? Borel attempts to answer the first question. Indeed, if the length of an interval is taken to be its measure, then according to (2), the measure of an open set is defined as the sum of lengths of the constituent intervals. Based on (3), the measure of a closed set is then defined (as the difference between an interval and an open set). Clearly, by using addition and subtraction in this manner one can obtain more and more complex sets, whose measure is defined according to (2) and (3).

This reasoning, given by Borel, should be viewed only as preliminary arguments. Borel did not present a mathematically rigorous description of the class of these sets, and, moreover, he did not show that postulate (1) will always be satisfied. These deficiencies were made up by Lebesgue who gave a rigorous construction of the class of sets obtained by successive application of addition and subtraction starting with open sets; this construction utilizes transfinite induction (see Lebesgue [6]). Lebesgue proposes that we call these sets B-measurable; they are also called Borel sets or B-sets. They are Lebesgue-measurable and their Lebesgue measure agrees with the measure defined using Borel's method.[1] Since the Lebesgue measure is always nonnegative, it follows that postulate (1) of Borel's measure is satisfied. [Postulate (1) indicates that the Borel measure is noncontradictory; in other words, the Borel measure of a set is independent of the manner in which this set is constructed; this fact is proved, however, using the methods of Lebesgue's theory (see Lusin [5]).]

In conclusion we present a translation of a proposition given in Borel's Leçons (p. 48) which was probably used by Lebesgue in the course of his definition of measure: "If a set E contains all the elements of a measurable set E_1 of measure α, then we may say that the measure E is larger than α without concern as to whether E is measurable or not.

---

[1] The shortest rigorous definition of the class of B-sets is as follows: the B-sets are the elements of the smallest σ-ring of sets containing the open sets. This definition is not constructive. Details on B-sets can be found in P. Halmos, "Measure Theory," Van Nostrand, Princeton, New Jersey, 1950.

Similarly, if E_1 contains all the elements of E, we say that the measure E is less than α. The terms greater or less do not exclude the case of equality . . ."

We shall see in the following chapters that this remark of Borel's had its share in the history of measure theory. In particular, it served as a weapon in the controversy between Borel [8]–[10] and Lebesgue [13], [14].[2]

[2] For additional details on Borel's ideas on measure and integration, see an extremely interesting and revealing recent paper of M. Fréchet [3]. (*Translator's note.*)

4 LEBESGUE'S MEASURE AND INTEGRATION

4.1 THE PROBLEM OF INTEGRATION

The first report by Lebesgue [1] about the discovery of a new integration process appeared in print in the spring of 1901 in *Comptes Rendus*; a detailed exposition of the theory of the integral was contained in his thesis "Intégrale, Longueur, Aire" [2] published in 1902. (From now on, we shall refer to this paper as ILA.) This theory, together with the analysis of the previous developments of the notion of the integral, constituted the content of his "Leçons sur l'Intégration et la Recherche des Fonctions Primitives," 1904 (98 of 138 pages were devoted to history). In what follows, we shall denote the first edition of this work as Leçons I, (Lebesgue [3]), and the second as Leçons II (Lebesgue [4]). (The second edition is markedly different from the first and contains new chapters.) These Leçons I are based on the course given by Lebesgue in the Collège de France during the 1902–1903 academic year. In Leçons I, his theory is presented in a more complete manner than in his thesis.

We shall base our exposition on Leçons I. Some special features of Lebesgue's thesis will be discussed later.

Analyzing the previous development of the integral, Lebesgue singles out its basic properties which determine the applicability of integrals as an instrument of mathematical analysis. This brought Lebesgue to the realization that these properties actually characterize integrals; if so, then one must investigate whether these properties can serve as a definition of the integral and how general a definition can be obtained.

Thus in Lebesgue's words: "it is our purpose to associate with every bounded function which is defined in a finite interval (a, b)—positive negative, or equal to zero—a certain finite number $\int_a^b f(x)\,dx$ which we will call the integral of f on (a, b) and which satisfies the following conditions:

(1)   For any a, b, and h, we have
$$\int_a^b f(x)\,dx = \int_{a-h}^{b-h} f(x + h)\,dx.$$

(2)   For any a, b, c we have
$$\int_a^b + \int_b^c + \int_c^a = 0.$$

(3) $\int_a^b [f(x) + \varphi(x)]\,dx = \int_a^b f(x)\,dx + \int_a^b \varphi(x)\,dx.$

(4)   If $f \geq 0$ and $b > a$, then also
$$\int_a^b f(x)\,dx \geq 0.$$

(5) $\int_0^1 1\,dx = 1.$

(6)   If f_n tends increasingly to f, then the integral of f_n tends to the integral of f.

The significance, necessity, and corollaries of the first five conditions of this problem of integration are more or less evident..." (Lebesgue [1], p. 98).

The whole of these six conditions represents a descriptive definition of an integral (see pp. 13, 14). We shall not dwell on the intrinsic characteristics of this descriptive definition suggested by Lebesgue[1] but we shall see how, according to Lebesgue, the required number $\int_a^b f(x)\,dx$ is determined.

The characteristic feature of all the Lebesgue arguments, as given in the following text, is the assumption that such a number does exist; under this assumption its properties are derived which, in the final analysis, will suggest a constructive definition of the integral.

We shall consider the case of a degenerated interval, i.e., the case in which, following Lebesgue's terminology, "the interval is equal to zero." If in Condition (2) we put $a = b = c$, we obtain that an integral over a degenerate interval is equal to zero. (We shall note for future reference that the integral over a degenerate interval is not identical with the integral over a set consisting of one point, see p. 61.)

We now show that

$$\int_a^b 0 \cdot dx = 0.$$

Indeed, if we put in Condition (3) $\varphi = f = 0$, then

$$\int_a^b (0 + 0)\,dx = \int_a^b 0 \cdot dx + \int_a^b 0 \cdot dx,$$

from which the required equality follows.

Furthermore, if we put in Condition (3) $\varphi = -f$, then

$$\int_a^b -f(x)\,dx = - \int_a^b f(x)\,dx. \tag{4.1}$$

We now show that for $f \leqslant \varphi$ and $a < b$,

$$\int_a^b f\,dx \leqslant \int_a^b \varphi\,dx. \tag{4.2}$$

---

[1] Some interesting remarks in connection with Conditions (1)–(6) are given in Lusin [4], [6] and in Leçons II (Lebesque [4]). Banach [1] has shown (using Zermelo's axiom) that there exists a number $\int_a^b f(x)$ that is defined for all bounded functions satisfying Conditions (1)–(5) and that does not satisfy Condition (6). [This means, in particular, that Condition (6) is not a corollary of Condition (1)–(5).]

For this purpose it is sufficient to note, using Conditions (4) and (3) and Eq. (4.1), that

$$0 \leqslant \int_a^b (\varphi - f)\, dx = \int_a^b \varphi\, dx + \int_a^b (-f)\, dx = \int_a^b \varphi\, dx - \int_a^b f\, dx.$$

From the inequalities $-|f| \leqslant f \leqslant |f|$ and from (4.2) and (4.1) we obtain

$$\left| \int_a^b f\, dx \right| \leqslant \int_a^b |f|\, dx.$$

Furthermore, from Condition (3) we obtain (n being an integer)

$$\int_a^b nf\, dx = n \int_a^b f\, dx,$$

and applying the last equality to f/n,

$$\frac{1}{n} \int_a^b f\, dx = \int_a^b \frac{1}{n} f\, dx.$$

From the last two equalities and from (4.1) we obtain

$$\int_a^b kf\, dx = k \int_a^b f\, dx \qquad (4.3)$$

for a rational k. Now let k be irrational and k_1 be rational. Then

$$\left| \int kf\, dx - k_1 \int f\, dx \right| < \int |k - k_1|\, |f|\, dx < |k - k_1|(\sup|f| + \eta) \int dx,$$

where η $(0 < \eta < 1)$ is chosen such that $|k - k_1|\,(\sup|f| + \eta)$ will be rational. If k_1 approaches k in this inequality, we obtain (4.3) also in the case k.

The reader is invited to show that

$$\int_a^b dx = c - a \qquad (4.4)$$

[here Conditions (1) and (5) should also be used].

Properties (4.2)–(4.4) were obtained without utilizing Lebesgue's condition (6). Therefore, the following important conclusion can be made: for functions integrable in the Riemann sense, the solution of the integration problem which satisfies Conditions (1)–(5) necessarily agrees with the Riemann integral. Indeed, subdividing $[a, b]$ into the intervals Δx_i and utilizing Condition (2) and properties (4.2)–(4.4) we obtain

$$\sum m_i \, \Delta x_i \leqslant \int_a^b f \, dx \leqslant \sum M_i \, \Delta x_i,$$

from which our conclusion follows.

Given a bounded function f on $[a, b]$, let $l < f(x) < L$ and $l = l_0 < l_1 < \cdots < l_n = L$; furthermore, let ψ_i be the characteristic function of the set $E_i = \{x \mid l_{i-1} < f(x) < l_i\}$ $(i = 1, \ldots, n)$. Then

$$\psi(x) = \sum_{i=1}^{n} l_{i-1} \psi_i(x) \leqslant f(x) \leqslant \sum_{i=1}^{n} l_i \psi_i(x) = \Psi(x).$$

It follows from this that

$$\int_a^b \psi(x) \, dx = \sum_{i=1}^{n} l_{i-1} \int_a^b \psi_i(x) \, dx \leqslant \int_a^b f(x) \, dx$$

$$\leqslant \sum_{i=1}^{n} l_i \int_a^b \psi_i(x) \, dx = \int_a^b \Psi(x) \, dx. \qquad (4.5)$$

The inequality (4.5) leads to a very important conclusion: the integral $\int_a^b f(x) \, dx$ is necessarily defined if the integrals of the functions ψ_i are defined. Indeed in this case, the integrals of the functions ψ and Ψ are defined and since the functions ψ and Ψ (which depend on the numbers $l_0, \ldots, l_n$) converge uniformly to the function f as $\max(l_i - l_{i-1}) \to 0$, the integral of f will necessarily be the common limit of the integrals of ψ and Ψ.[2]

Thus, Lebesgue, consistently searching for the conclusions which should necessarily follow from the Conditions (1)–(6) of the integration problem, arrives at the following result: it is only necessary to be able to define an integral for functions admitting two values 0 and 1; if this is accomplished then the integral of any bounded function, provided it exists, will necessarily be the limit of integrals of step functions (4.5).

---

[2] Uniformly convergent sequences can be integrated term by term: if f_n converges uniformly to f, then

$$\left| \int_a^b f_n \, dx - \int_a^b f \, dx \right| \leqslant \int_a^b |f_n - f| \, dx \leqslant (b - a) \sup_{[a, b]} |f_n - f| \to 0.$$

Thus the general problem of integration is reduced to the simpler problem of integrating characteristic functions.

The characteristic function ψ_E of the set E is completely determined by the set E and hence, $\int_a^b \psi_E(x) \, dx$ is a number which depends on the part of the set E belonging to $[a, b]$. When E is the segment $[a, b]$, this number is equal to the length $b - a$ of the segment [see (4.4)]; therefore it is natural in the general case to call this number the measure of the set E. The problem of integration of characteristic functions is thus reduced to the measure problem. We now proceed to a consideration of Lebesgue's solution of this problem.

In Lebesgue's words: "Here is the problem which is to be solved: Our purpose is to associate with every bounded set E consisting of points on the x axis a certain nonnegative number, mE, which will be called the measure of E, and which satisfies the following conditions:

(1′)   Two congruent sets have the same measure.

(2′)   A set which is the sum of a finite or countable number of pairwise disjoint sets has a measure equal to the sum of the measures of the summands.

(3′)   The measure of the set of all points of the interval $(0, 1)$ equals $1 \ldots$ " (Lebesgue [1], p. 103).

*We shall show that Conditions (1′)–(3′) actually correspond to Conditions (1)–(6) of the problem of integration, formulated for characteristic functions.

Condition (1′) is equivalent to Condition (1): the function $\psi_E(x + h)$ is the characteristic function of the set obtained from E by a translation to the left in the amount h of the whole axis ($[a, b]$ is thus shifted to $[a - h, b - h]$), and Condition (1) expresses the fact that the measure is translation invariant, i.e., congruent sets have the same measure.

Condition (3′) clearly corresponds to Condition (5).*

Condition (2) requires that the measure of a set contained in $[a, b]$ be equal to the sum of the measures of its parts belonging to $[a, c]$ and $[c, b]$ (the point c may possibly be counted twice). We shall see that, formulated in this manner, Condition (2) is a corollary of Conditions (3) and (4).

Condition (3) makes sense for the measure problem if and only if f and φ are characteristic functions of nonoverlapping sets, since only in this case is $f + \varphi$ a characteristic function (of the sum of these sets). In this case, Condition (3) is the condition of *finite additivity of the measure*:

$$m(E_1 + E_2) = mE_1 + mE_2, \qquad E_1 E_2 = 0.$$

According to Condition (4), the measure is nonnegative; this together with the finite additivity implies monotonicity: from $E_1 \subset E_2$ it follows that $mE_1 \geqslant mE_2$. Monotonicity, additivity, and congruence imply that the measure of a one-point set $\{a\}$ equals zero, since if the points $a_1, \ldots, a_n$ belong to $[0, 1]$, then $m\{a_1\} + \cdots + m\{a_n\} \leqslant 1$, and since n is arbitrary, $m\{a_1\} = \cdots = m\{a_n\} = 0$. Thus, Condition (2) indeed follows from Conditions (3) and (4).

Turning now to Condition (6) we note that the monotonicity of a sequence of characteristic functions $\{\psi_{\mathcal{E}_n}\}$ implies the monotonicity of the sequence of sets $\{\mathcal{E}_n\}$; from Condition (6) we should have

$$\lim_{n \to \infty} m\mathcal{E}_n = m\left(\lim_{n \to \infty} \mathcal{E}_n\right). \tag{4.6}$$

Condition (4.6) is equivalent to the condition of *countable additivity*. Indeed, if $\{E_i\}$ is a sequence of nonoverlapping sets, $E_i E_j = 0$, then putting $\mathcal{E}_n = \sum_1^n E_i$ and utilizing finite additivity we obtain Condition (2′):

$$\sum_i mE_i = m \sum_i E_i.$$

Thus, Condition (2′) implies Conditions (2)–(4) reformulated for the case of characteristic functions, and Conditions (1)–(5) of the integration problem reduce for this case to Conditions (1′)–(3′).*

4.2   THE MEASURE PROBLEM

We have seen that the measure of the segment $[a, b]$ [or of the interval (a, b), since the measure of a point is zero] equals $b - a$. In view of Condition (2′) the measure of an open set A is equal to the sum of the lengths of the intervals whose union is A. (As before, the characteristic feature of our arguments is that the existence of the measure is assumed.)

Now let E be an arbitrary set with no information about its structure. Let G be an open set containing E. If mE is defined, then in view of the monotonicity of the measure, $mE < mG$ and hence, $mE \leqslant \inf mG$, where the *inf* is taken over all the open sets containing E. The number $\inf mG$ is the required upper bound on mE. Lebesgue puts $\inf mG = m_e E$ and calls $m_e E$ the *outer measure* of the set E. We now bound mE from below. We have $mE + mCE = b - a$, $mE = (b - a) - mCE$, and hence, $mE \geqslant (b - a) - m_e CE$. The lower bound on mE is then obtained. Lebesgue calls the number $m_i E = b - a - m_e CE$ the *inner measure* of the set E. The numbers $m_e E$ and $m_i E$ are defined for an arbitrary set irrespective of whether or not the measure mE exists. (Starting from this point the numbers $m_i E$, $m_e E$, and mE denote respectively, the inner measure, outer measure, and Lebesgue measure of the set E; for the PJ-measure the corresponding notation will be $e_i E$, $e_e E$, and eE). We find the relation between $m_i E$ and $m_e E$. For this purpose, consider open sets G_1 and G_2 containing E and CE, respectively. Then we have $mG_1 + mG_2 \geqslant b - a$.[3] Taking the infimum of the terms in the left-hand side of the inequality we obtain

$$m_e E + m_e CE \geqslant b - a \qquad \text{or} \qquad m_e E \geqslant m_i E.$$

Incidentally, it follows from the definitions of the outer and the inner Peano–Jordan measures that

$$e_i E \leqslant m_i E \leqslant m_e E \leqslant e_e E. \tag{4.7}$$

Hence, if the solution of the measure problem exists, then for PJ-measurable sets, mE coincides with the PJ-measure.

We now assume that "luckily" we obtain the equality $m_i E = m_e E$ for some E; then the common value of the outer and inner measures should be equal to the measure of E (if it exists).

This remark is the ultimate result in the analysis of the integration problem; it prompted Lebesgue to come out with the following constructive definition of a measure.

Definition 4.1

The set E is said to be measurable if $m_i E = m_e E$, and this common value is then called the Lebesgue measure of E.

---

[3] If the system of intervals contains the segment $[a, b]$, then in view of Borel's lemma there exists a finite number of intervals whose sum contains $[a, b]$; the sum of their lengths will clearly be no less than $b - a$.

Lebesgue confines himself to the investigation of measure only for these sets (without asserting however that a solution of the measure problem for a wider class of sets does not exist; see, e.g., Natanson [1]).

To verify that the number mE satisfies conditions (1')–(3') we formulate the measurability criterion.[4]

Measurability Criterion

In order that a set $E \subset [a, b]$ be measurable it is necessary and sufficient that for every $\varepsilon > 0$ there exist open sets G_1, G_2 such that $G_1 \supset E$, $G_2 \supset CE$, and $m(G_1 \cdot G_2) < \varepsilon$.

The proofs of this criterion and of the following assertions are simplified by application of the lemma: *If G_1, G_2 are open subsets of $[a, b]$ given that $m(G_1, G_2) < \varepsilon$, then*

$$mG_1 + mG_2 \leqslant b - a + \varepsilon.$$

We shall not prove the criterion and the lemma here: neither shall we verify that Conditions (1') and (3') are satisfied by the Lebesgue measure (the proof of the latter is rather straightforward).

We thus turn to the verification of Condition (2').

Verification of Condition (2')

First we show that if $G = \sum_i G_i$ then

$$mG \leqslant \sum_i mG_i, \qquad (\alpha)$$

where G, G_i are open sets. Indeed, first let G be a single interval; we take a segment $\Delta \subset G$ and apply Borel's lemma to the system of intervals of the sets G_i, covering Δ. We thus obtain $m\Delta \leqslant \sum_i mG_i$. Since Δ is an arbitrary segment contained in the interval G, inequality (α) is proved. The same argument holds if G consists of a finite number of intervals; the general case when $G = \sum_{i=1}^{\infty} \delta_i$, is obtained from the limiting process as $n \to \infty$ in the inequality $\sum_{i=1}^{n} \delta_i \leqslant \sum_i mG_i$.

---

[4] We emphasize once more that the original arguments of Lebesque's measure theory are arguments concerning systems of intervals utilizing the notion of the limit of a monotonic sequence, the notion of the infimum of a set of real numbers and nothing else. In the following text, the term "open set" will denote a system of nonoverlapping open intervals; mG denotes the sum of the lengths of the intervals comprising the set G.

Now let $E \subset \sum_i E_i$. Define open sets $G_i \supset E_i$ such that

$$mG_i \leqslant m_e E_i + 2^{-i} \cdot \varepsilon.$$

Since $E \subset \sum G_i$, then as we have just proved,

$$m_e E \leqslant \sum mG_i \leqslant \sum m_e E_i + \varepsilon,$$

i.e.,

$$m_e E \leqslant \sum m_e E_i. \tag{4.8}$$

Let $G = \sum_i \delta_i$ and $F = CG$. A simple argument shows that it is possible to define an open set $\tilde{G}$, $\tilde{G} \supset F$, such that

$$m\tilde{G} + mG \leqslant b - a + \varepsilon, \qquad m(G \cdot \tilde{G}) < \varepsilon.$$

These inequalities will be utilized in the following text; in particular the second inequality which proves (in view of the measurability criterion) the measurability of open sets and their complementary closed sets.

Now let E_1, E_2 be two arbitrary nonoverlapping sets on the segment $[a, b]$; put $E_3 = E_1 + E_2$. Define open sets G_i, $i = 1, 2, 3$ such that (a) $G_i \supset CE_i$; (b) $mG_i \leqslant m_e CE_i + \varepsilon$; (c) $G_1 \supset G_3$, $G_2 \supset G_3$. Let $F_i = CG_i$, $i = 1, 2$; clearly, $F_i \subset E_i$ and therefore, (d) $F_1 \cdot F_2 = 0$. As was stated above, there exist open sets $\tilde{G}_i \supset F_i$, $i = 1, 2, 3$ such that (e) $m(\tilde{G}_i \cdot G_i) < \varepsilon$; in view of (c) we can assume that (f) $\tilde{G}_1 \cdot \tilde{G}_2 = 0$. Since $\tilde{G}_i + G_i \supset [a, b]$, we have

$$m\tilde{G}_1 + mG_1 \geqslant b - a,$$
$$m\tilde{G}_2 + mG_2 \geqslant b - a.$$

Adding up these inequalities we obtain

$$m\tilde{G}_1 + m\tilde{G}_2 \geqslant (b - a) - mG_1 + (b - a) - mG_2. \tag{β}$$

Utilizing (c), (e), and (f) and the lemma stated above we obtain

$$m\tilde{G}_1 + m\tilde{G}_2 + mG_3 \leqslant b - a + 2\varepsilon.$$

Combining this inequality with inequality (β), we finally obtain in view of (a) and (c) that

$$m_i E_3 \geqslant m_i E_1 + m_i E_2.$$

By induction we obtain that for each n

$$m_i \sum_{k=1}^{n} E_k \geqslant \sum_{k=1}^{n} m_i E_k,$$

and sending n to infinity we have for the case of a countable number of disjoint sets

$$m_i \sum_{1}^{\infty} E_k \geqslant \sum_{1}^{\infty} m_i E_k. \tag{4.9}$$

Combining (4.8) and (4.9) we obtain

$$m \sum E_k = \sum m E_k,$$

provided the sets E_k are measurable. We thus have shown that *a finite or countable sum of pairwise disjoint measurable sets is measurable and, moreover, Condition (2′) is satisfied.*

We finally show the measurability of the sum of measurable, possibly overlapping, sets. Consider first the case of a finite number of summands: let E_1, E_2 be measurable sets and let $E_3 = E_1 + E_2$. Define the open sets $G_1 \supset E_1$, $G_2 \supset E_2$, $\tilde{G}_1 \supset CE_1$, $\tilde{G}_2 \supset CE_2$ such that $m(G_1 \cdot \tilde{G}_1) < \varepsilon$ and $m(G_2 \cdot \tilde{G}_2) < \varepsilon$. Let $G_3 = G_1 + G_2$, $\tilde{G}_3 = \tilde{G}_1 \cdot \tilde{G}_2$. Then, clearly, $G_3 \supset E_3$, $\tilde{G}_3 \supset CE_3$, and $m(G_3 \cdot \tilde{G}_3) < 2\varepsilon$. In view of the measurability criterion the sum $E_1 + E_2$ is measurable. The general case $E = \sum_{1}^{\infty} E_k$ of a countable number of terms is reduced to the cases just considered by writing $E = E_1 + \sum_{2}^{\infty} [E_i - (E_{i-1} + \cdots + E_1)]$.

We also note that following directly from the definition of measurability, the sets E and CE are measurable simultaneously (namely if E is measurable so is CE and conversely). We then can summarize Lebesgue's results as follows: *The collection of Lebesgue measurable (L-measurable) sets forms a σ-ring on which the Lebesgue measure is a solution of the measure problem.*

As we have noted above, open sets are measurable; also the operations of addition and subtraction, which define the B-sets leave us within the bounds of a σ-ring, hence *the B-sets are Lebesgue measurable and their measure as defined by Borel agrees with the Lebesgue measure.* In particular, the consistency of the Borel measure follows from here (cf. footnote 1, p. 46).

The class of L-measurable sets is wider than the class of B-measurable sets. This, as Lebesgue shows, can be seen by a simple comparison of their cardinalities; on the one hand, there is only a continuum of B-sets[5]; on the other hand, all subsets of a perfect set of measure zero are also L-measurable (and of measure zero); the cardinality of the set of these subsets is equal to the cardinality of the set of all sets on the real line and hence is higher than the cardinality of the continuum.

In his thesis [2] Lebesgue observes that for every set E there are two B-sets $B_1(E)$ and $B_2(E)$ such that $B_1(E) \subset E \subset B_2(E)$ and $mB_1(E) = m_iE, mB_2(E) = m_e E.$ This fact follows from the definition of the outer measure: we can define a sequence of open sets $G_n \supset E$ such that $mG_n < m_e E + 1/n$ and put $B_2(E) = \prod_1^\infty G_n.$ Analogously one can construct a set $B_2(CE)$ for the complement CE and put $B_1(E) = [a, b] - B_2(CE).$

4.3 MEASURABLE FUNCTIONS

Following Lebesgue, we now return to the integration problem considered in Section 4.1. We have seen that the problem of defining an integral of an arbitrary function bounded on the given segment $[a, b]$ can be solved by using inequality (4.5) if the integrals of characteristic functions are defined. However, the integration problem of characteristic functions was solved by Lebesgue not for arbitrary functions of this class but *only for a class of functions which are characteristic functions of L-measurable sets.* Thus the class of functions which are involved in the inequality (4.5) is narrowed down. If we require that characteristic functions of *measurable* sets appear in (4.5), we must then consider only those functions f for which the sets of the form $E_x(\alpha \leqslant f < \beta)$ are measurable. Lebesgue calls a function f for which the sets $E_x(\alpha \leqslant f < \beta)$ are measurable for arbitrary real α and β, a *measurable function*. We shall note immediately that measurable functions can be defined by requiring the measurability of sets of the form $E_x(f > \alpha)$ or of the form

---

[5] Indeed, there is a continuum of at most countable systems of intervals with rational end points. Since each B-set can be obtained by means of a finite or countable number of operations performed over these intervals, it follows that there is a continuum of B-sets.

$E_x(f \geqslant \alpha)$, or of their complements. The equivalence of the corresponding definitions follows from the properties of σ-rings of sets.

Continuous functions are measurable since for such functions the sets $E_x(f \geqslant \alpha)$ are closed. It also follows from the properties of the class of L-measurable sets that the sum of two measurable functions is measurable and the limit of a sequence of measurable functions is measurable; in particular the limit of a sequence of continuous functions is measurable, i.e., the functions of the first Baire class are measurable; for the same reason, any function in the Baire classification is measurable.[6] The definition of measurable functions can also be used without any change when the domain of definition of the functions is an arbitrary set E. A function defined on a set of measure zero is measurable; if a function is measurable on the sets E_1 and E_2 separately, then it is measurable on their sum $E_1 + E_2$. Let f be an R-integrable function on $[a, b]$ and let E_1 be the set of its discontinuity points and let $E_2 = [a, b] - E_1$. Then $mE_1 = 0$ (from the Riemann integrability criterion), and since $[a, b] = E_1 + E_2$, and, moreover, f is a continuous function on E_2 (and hence measurable), it follows that every R-integrable function is measurable.

4.4 AN ANALYTICAL DEFINITION OF THE INTEGRAL

The definition of an integral of a bounded measurable function f as the common limit of sums

$$
\begin{aligned}
\sigma &= \sum_{i=0}^{n-1} l_i \left\{ mE_x (f(x) = l_i) + mE_x (l_i < f(x) < l_{i+1}) \right\}, \\
\Sigma &= \sum_{i=1}^{n} l_i \left\{ mE_x (l_{i-1} < f(x) < l_i) + mE_x (f(x) = l_i) \right\},
\end{aligned}
\tag{4.5'}
$$

is termed by Lebesgue the *analytical definition*. These sums are essentially the same as those appearing in the inequalities (4.5). The conventional proof of the existence of this limit is in essence the proof given by Lebesgue, and will not be reproduced here. All that remains is to verify

---

[6] These are all the functions (and only those) for which the sets $E_x(f > \alpha)$ are Borel-measurable. A discussion of the Baire classification can be found, for example, in Natanson [1].

that an integral defined in this manner satisfies properties (1)–(6) of the integration problem; Lebesgue proves this in detail for Conditions (3) and (6). He first observes that property (3) is obvious for linear combinations of characteristic functions [of the form of functions ψ and Ψ in the inequality (4.5)]; the general case is obtained by taking the limit; this limiting process is permissible in view of the definition of the integral.

Condition (6) is verified in its well-known form: if the functions f_n are uniformly bounded and f_n converges to f, then the integrals of f_n converge to the integral of f. The proof is based on the fact that $\lim mE_n = 0$, where $E_n = E_x(|f(x) - f_n(x)| < \varepsilon)$; this fact is not proved by Lebesgue however. It seems that the following theorem formulated by Lebesgue [7] in 1903 is utilized here: *if there exists a convergent series of measurable functions* $\sum f_n = f$, *then*

$$\lim_{N \to \infty} mE\left(\sup_{x} \left(\sup_{n > N}|\Phi_n - f| > \varepsilon\right)\right) = 0,$$

where $\Phi_n \overset{\text{def}}{=} \sum_1^n f_i$ (compare with Lebesgue [8], *Introduction*, where this theorem is proved).

The discussions above were concerned with the functions defined on $[a, b]$; in the case when f is given on a set $E \subset [a, b]$, Lebesgue suggests that we *complete its definition* by assuming the value 0 on $[a, b] - E$ and calling it integrable (see below) if the function f_1 resulting from this extension is integrable on $[a, b]$. Moreover,

$$\int_E f \, dx \overset{\text{def}}{=} \int_a^b f_1 \, dx.$$

4.5 INTEGRABLE (SUMMABLE) FUNCTIONS

Lebesgue observes that the preceding definitions of the integral are applicable to certain unbounded functions, namely those for which at least one of the series

$$\sigma = \sum_{-\infty}^{+\infty} l_i \, mE_x (l_i \leqslant f(x) < l_{i+1}),$$

$$\Sigma = \sum_{-\infty}^{+\infty} l_i \, mE_x (l_{i-1} < f(x) \leqslant l_i),$$

(4.10)

is convergent for a certain choice of numbers l_i, satisfying conditions:

$$\lim_{i \to +\infty} l_i = +\infty, \qquad \lim_{i \to -\infty} l_i = -\infty, \qquad l_{i+1} - l_i < \alpha \quad (\alpha > 0)$$

for every i. However in this case the convergence of one series implies the convergence of the other for any l_i with uniformly bounded difference $l_{i+1} - l_i$.[7] In the usual manner, it is proved that in this case the sums σ and Σ converge to the same limit as $\alpha \to 0$. These functions are called by Lebesgue *summable* (integrable) *functions*.

As an example of a function that is not summable, Lebesgue considers the function

$$y(x) = \left(x^2 \sin \frac{1}{x^2}\right)' = 2x \sin \frac{1}{x^2} - \frac{2}{x} \cos \frac{1}{x^2}; \qquad y(0) = 0. \quad (4.11)$$

The nonsummability of $y(x)$ follows from the summability (integrability) of the first term and the nonsummability of the second term in the right-hand side of (4.11). (This function will be referred to again in Section 8.1)

The function (4.11) is integrable in the improper Cauchy sense (or in the Harnack sense). Lebesgue points out that the Cauchy and Dirichlet methods are applicable for the definition of integrals of nonsummable functions; however in Leçons I he does not treat this topic in great detail.

Lebesgue's definition of the integral of a summable function is presently one of several well-known definitions. In Leçons II, Lebesgue

---

[7] Indeed, let, for example, the series σ be convergent. Consider a sequence of numbers $\{k_i\}$, $k_i \to \pm \infty$, for $i \to \pm \infty$, $0 < k_{i+1} - k_i < \alpha$, different from the sequence $\{l_i\}$. Let $E_i \overset{\text{def}}{=} E_x(l_{i-1} \leqslant f < l_i)$, $\mathscr{E}_i \overset{\text{def}}{=} E_x(k_{i-1} \leqslant f < k_i)$, $\mathscr{E}_{ij} \overset{\text{def}}{=} E_i \cdot \mathscr{E}_j$. Then one can write $E_i = \sum_{j=m(i)}^{j=n(i)} E_i \mathscr{E}_j$; $m(i)$ and $n(i)$ are such that $E_i \cdot \mathscr{E}_m \neq 0$, $E_i \cdot \mathscr{E}_n \neq 0$, $E_i \cdot \mathscr{E}_{m-p} = E_i \cdot \mathscr{E}_{n+p} = 0$ for $p = 1, 2, \ldots$, and

$$\sum_{-\infty}^{+\infty} l_i m E_i = \sum_{-\infty}^{+\infty} l_i \sum_{j=m(i)}^{j=n(i)} m\mathscr{E}_{ij} = \sum_{j=-\infty}^{\infty} \sum_{i=m_1(j)}^{i=n_1(j)} l_i m\mathscr{E}_{ij}.$$

It is, however, evident that $k_{j-1} < l_{m(j)} \leqslant k_{j+1}$ and, hence,

$$k_{j-1} \sum_{m_1(j)}^{n_1(j)} m\mathscr{E}_{ij} \leqslant \sum_{m_1(j)}^{n_1(j)} l_i m\mathscr{E}_{ij} \leqslant k_{j+1} \sum_{m_1(j)}^{n_1(j)} m\mathscr{E}_{ij}.$$

From here we have

$$\sum_{m_1(j)}^{n_1(j)} l_i m\mathscr{E}_{ij} = (k_j + \theta\varepsilon_1) m\mathscr{E}_j, 0 \leqslant |\theta| \leqslant 2.$$

Finally, $\sum l_i m E_i = \sum k_j m\mathscr{E}_j + \theta\varepsilon_1 m E$ which proves the convergence of the series $\Sigma = \sum k_j m\mathscr{E}_j$.

proves that the integrals of truncated functions $f_M{}^N(x)$ converge to the integral of the function f. Hence, it is possible to define an integral of a summable function using the de la Vallée-Poussin procedure; this is the method used in many textbooks. We shall come across some other methods of defining L-integrals in the text.

The above discussion is about all that was said by Lebesgue (in Leçons I) in connection with the definition of summability. In Leçons II (which contains 342 pages as compared to 138 pages in Leçons I), there is a more detailed analysis of this definition. The following properties completely characterize the integral of a summable function on E. (These conditions were formulated by Lebesgue [12] in 1910.)

(1) If E_h is the set of those values of x for which $x - h$ belongs to E, then

$$\int_E f(x)\, dx = \int_{E_h} f(x - h)\, dx.$$

(2) If $E = \sum_{0 \leqslant i \leqslant \infty} E_i$, $E_i \cdot E_j = 0$, then

$$\int_E f(x)\, dx = \sum_{0 \leqslant i \leqslant \infty} \int_{E_i} f(x)\, dx.$$

(3) $\int_E (f(x) + \varphi(x))\, dx = \int_E f(x)\, dx + \int_E \varphi(x)\, dx.$

(4) If $f(x) \geqslant 0$, then

$$\int_E f(x)\, dx \geqslant 0.$$

(5) $\int_0^1 1\, dx = 1.$

Here requirement (6) of the integration problem is absent, however requirement (2) is sharpened.

It is easy to see that these five requirements actually define measurable sets and summable functions together with their integrals. Indeed, for characteristic functions requirements (1), (2), and (5) determine the measure problem discussed in Section 4.2, i.e., these requirements determine, as above, the measurable bounded functions and their integrals. Condition (2) uniquely defines an integral of an unbounded (measurable) function: if we take two sequences of numbers $\{k_i\}$ and $\{l_i\}$, the first of which monotonically decreases to $-\infty$, and the second monotonically increases to $+\infty$, (with $k_1 = l_1$) and put $E_i = E_x(k_{i+1} \leqslant f(x) < k_i) + E_x(l_i \leqslant f(x) < l_{i+1})$, then we have in view of (2),

$$\int_E f(x)\, dx = \lim_{n \to \infty} \sum_1^n \int_{E_i} f(x)\, dx = \lim_{i \to \infty} \int_{\substack{E(k_{i+1} \leqslant f(x) < l_{i+1}) \\ x}} f(x)\, dx,$$

for any sequence $\{k_i\}$ and $\{l_i\}$. The left-hand side of this equality is defined as the limit of the right-hand side; this definition is a slight modification of the de la Vallée-Poussin procedure.

4.6 A GEOMETRICAL DEFINITION OF THE INTEGRAL

We are now familiar with the geometrical definition of the Riemann integral in which the integral is interpreted as the Jordan measure of an ordinate set. An analogous definition of the integral was also suggested by Lebesgue. We shall now give a proof of the equivalence between the analytical and geometrical definitions of the integral of a bounded function in the form presented in Leçons I. We shall use the notion of measurability of a planar set without any further clarification; the theory of L-measurable sets in the plane is analogous to that on the line. (See, e.g., Natanson [1], W. Rudin, "Real and Complex Analysis," McGraw-Hill, New York, 1966, or P. Halmos, "Measure Theory," Van Nostrand, Princeton, New Jersey, 1950.)

In this section, we denote by $m_s E$, $m_{s,\,i} E$, and $m_{s,\,e} E$, respectively, the measure, inner measure, and outer measure of a planar set E; the same quantities for a linear set will be denoted by $m_l E$, $m_{l,\,i} E$, and $m_{l,\,e} E$, respectively. We also define the ordinate sets

$$E^+(f) \overset{\text{def}}{=} E(f^+, [a, b]), \qquad E^=(f) \quad E(f^-, [a, b]).$$

*To simplify the exposition we now discuss several assertions concerning planar sets, which we shall use in the text.

Assertion 1

A bounded linear set has planar measure zero.

Proof

Every segment of the line can be included in a strip of arbitrarily small area.

Assertion 2

If an open set G is such that its horizontal sections $y = k$, $a \leqslant k \leqslant b$ have linear measure $> h$, then $m_s G > h(b - a)$.

To prove the assertion consider first the case where G is a finite sum of squares with sides parallel to the coordinate axis; the general case where G is a countable sum of such squares is obtained by taking the limit.

Assertion 3

If $\mathscr{E}$ is a measurable linear, then $E = E_{(x, y)}(x \in \mathscr{E}, 0 \leqslant y \leqslant b)$ is a measurable planar set and, moreover, $m_s E = b \cdot m_l \mathscr{E}$.

Proof

The assertion is obvious in the case when $\mathscr{E}$ is an interval or a segment; in view of the additivity of the measure it is also valid in the case when $\mathscr{E} = \sum \delta_i$ is an open set (a sum of the constituent intervals), since in this case it is true that

$$\underset{(x, y)}{E} (x \in \mathscr{E}, 0 \leqslant y \leqslant b) = \sum \underset{(x, y)}{E} (x \in \delta_i, 0 \leqslant y \leqslant b);$$

the assertion is valid furthermore in the case when $\mathscr{E}$ is a closed set (a complement of an open set relative to an interval). In the general case there exist a closed set $F \subset \mathscr{E}$ and an open set $G \supset \mathscr{E}$ such that

$$m_l G - m_l F < \varepsilon$$

and

$$\underset{(x, y)}{m_s E} (x \in G, \ 0 \leqslant y \leqslant b) - \underset{(x, y)}{m_s E} (x \in F, \ 0 \leqslant y \leqslant b) < (b - a)\varepsilon;$$

from here the assertion follows.*

The Geometrical Definition of the Integral

Let f be a bounded function on $[a, b]$. *Then*

$$\int_a^b f(x)\, dx \stackrel{\text{def}}{=} m_s E^+(f) - m_s E^-(f),$$

provided the right-hand side is well defined (i.e., when $E^+(f)$ *and* $E^-(f)$
are measurable sets).

Theorem 4.1

The geometric and analytic definitions of integrals of bounded functions are equivalent.

Proof

Let the sets E^+, E^- be measurable. To prove the measurability of f it is sufficient to prove the measurability of f^+ and f^-. Therefore, we can assume that $f(x) \geqslant 0$. The quantities $m_{l,\,i} E_x(f \geqslant \alpha) = m_i(\alpha)$, $m_{l,\,e} E_x(f \geqslant \alpha) = m_l(\alpha)$ are monotonically decreasing functions of the parameter α and, moreover, $m_i(\alpha)$ is left-continuous.[8] Assume that for some α'

$$m_l(\alpha') > m_i(\alpha') + \varepsilon.$$

It follows from the left continuity of $m_i(\alpha)$ and the monotonicity of $m_l(\alpha)$ that there exists an $h > 0$ such that

$$m_l(\alpha' - k) > m_i(\alpha' - k) + \frac{\varepsilon}{2}, \qquad 0 \leqslant k \leqslant h. \tag{4.12}$$

The set

$$E_h \stackrel{\text{def}}{=} \underset{(x,\,y)}{E}\, (a \leqslant x \leqslant b, \quad \alpha' - h \leqslant y \leqslant \alpha') \cdot E(f)$$

is measurable, being a product of two measurable sets; hence there exist open sets A and B each of which can be represented as a sum of a finite or countable number of mutually nonoverlapping squares and such that $A \supset E_h$, $B \supset CE_h$, where CE_h is the complement of E_h with respect to

---

[8] In both editions of Leçons, Lebesgue asserts the left-continuity of $m_l(\alpha)$ also; this does not follow from the properties of the measure and is not necessary for his arguments.

the rectangle containing the ordinate set, and $D \overset{\text{def}}{=} A \cdot B$ is a system of squares of total measure $\leqslant \eta$, where $\eta > 0$ is an arbitrarily small number. Furthermore, the intersection of A and B with the line $y = d$, $\alpha' - h \leqslant d \leqslant \alpha'$ is a system of intervals containing the sets (translated on the line $y = d$) $E_x(f \geqslant d)$ and $CE_x(f \geqslant d)$, respectively. The common part of this system of intervals is the intersection of D with the line $y = d$. The measure of this intersection is larger than $(\varepsilon/2)$ in view of (4.12) and Assertion 2 can be applied to D. We thus obtain $\eta \geqslant (\varepsilon/2) \cdot h$. But this is a contradiction, since ε and h are fixed and η can be chosen independently to be arbitrarily small. Thus we have proved that if the set $E(f)$ is measurable then the inequality (4.12) cannot hold; hence f is measurable and the analytical definition of the integral can be applied to it.

We now show that if f is measurable then $E(f)$ is measurable and $\int_a^b f \, dx = m_s E(f)$. This will complete the proof of the theorem.

Let $l_0 < f(x) < l_n$. Consider the sets

$$e_n \overset{\text{def}}{=} \sum_{i=1}^n \underset{(x, y)}{E} (l_{i-1} \leqslant f(x) < l_i, \quad 0 \leqslant y \leqslant l_{i-1}),$$

$$\mathscr{E}_n \overset{\text{def}}{=} \sum_{i=1}^n \underset{(x, y)}{E} (l_{i-1} \leqslant f(x) < l_i, \quad 0 \leqslant y \leqslant l_i).$$

In view of Assertion 3, e_n and $\mathscr{E}_n$ are measurable sets and, moreover, $e_n \subset E(f)$, $\mathscr{E}_n \supset E(f)$, and also $m_s \mathscr{E}_n - m_s e_n \to 0$ if $\max_i(l_i - l_{i-1}) \to 0$. Hence, $E(f)$ is measurable and

$$m_s E(f) = \lim m_s \mathscr{E}_n = \lim m_s e_n = \int_a^b f \, dx. \qquad \text{Q.E.D.}$$

Remark. For the case when the ordinate set is nonmeasurable, Lebesgue introduces the notions of upper and lower integrals defined as the outer and inner measures of the ordinate set if $f \geqslant 0$, and in the general case defined as $m_{s, e} E^+(f) - m_{s, i} E^-(f)$, and $m_{s, i} E^+(f) - m_{s, e} E^-(f)$, respectively.

4.7   LEBESGUE'S INTEGRAL AND THE PROBLEM OF THE PRIMITIVE

Following the exposition given in Leçons I, we shall now investigate to what extent the Lebesgue integration recovers the primitive function (see Section 2.15). Assume f' exists everywhere in the interval $[a, b]$.

The case of bounded derivatives is immediately solved by the theorem on term-by-term integration of a bounded sequence of functions. Indeed, in the equality

$$f'(x) = \lim_{h \to 0} r(f, x, x + h),$$

h can run through a certain countable sequence $\{h_n\}$ approaching zero; then $f'(x) = \lim r(f, x, x + h_n)$ where $r(f, x, x + h_n)$ are continuous functions of x,[9] and uniformly bounded [this follows from the mean value theorem: $r(f, x, x + h) = f'(x + \theta_n h_n)$]; therefore we can write

$$\int_a^x f'(x)\, dx = \lim \frac{1}{h_n} \left[\int_a^x f(x + h_n)\, dx - \int_a^x f(x)\, dx \right]$$

$$= \lim \frac{1}{h_n} \left[\int_{a+h_n}^{x+h_n} f(x)\, dx - \int_a^x f(x)\, dx \right]$$

$$= \lim \left[\frac{1}{h_n} \int_x^{x+h_n} f(x)\, dx - \frac{1}{h_n} \int_a^{a+h_n} f(x)\, dx \right]$$

$$= f(x) - f(a).$$

Thus, Lebesgue's integral recovers the primitive of a bounded derivative. The following general theorem is valid.

Theorem 4.2

The Lebesgue indefinite integral of a summable finite derivative is a primitive function.

In various points in Leçons I and particularly in the course of the proof of Theorem 4.2, Lebesgue uses the method he calls *the method of interval chains*. It is based on the following fact. Let there be a system of intervals A such that for each point $x \in [a, b)$ there exists at least one interval in A with left endpoint at x. Then there exists a system A_1 of intervals belonging to A such that each point in $[a, b)$ either belongs to a unique interval in A_1 or it is the left endpoint of such an interval. Such a system of intervals is called a *chain of intervals*; we say that

---

[9] From here, in particular, the measurability of the derivative f' and, moreover, its membership in the first Baire class follows.

a chain *connects a and b* if *b* does not belong to an interval in the chain; otherwise we say that the chain *covers* [*a, b*]. The length of the largest interval in the chain is called the *diameter of the chain.*

*Proof of the Existence of a Chain

The required intervals are arranged in a transfinite sequence as follows: The first interval is located to the right of the point *a*; assume that for some finite or transfinite γ intervals are constructed which form a chain and are enumerated from left to right by all finite or transfinite numbers smaller than γ. Consider the supremum of the right ends of these intervals, and to the right of it locate an interval which shall be designated by the number γ. Thus we have shown how an interval with an arbitrary finite or transfinite index is constructed to form a chain with the preceding intervals.

Every point in [*a, b*) is either an end point or an inner point of some interval with a finite or transfinite index of at most the second class; otherwise we would be able to locate in [*a, b*] an uncountable system of nonoverlapping intervals (enumerated by all the ordinals of the first and second classes) which is impossible. The proof is thus completed.[10]

In certain cases it is possible to prove the existence of a chain without using transfinite numbers.

A discussion of proofs using chains of intervals is contained in Leçons II.*

Remark. Assume that for each point $x \in$ [*a, b*] in the system *A* there exist *arbitrarily* small intervals with left endpoint at *x*; under this condition there exists a chain which *connects a* and *b*. Indeed, to construct the chain we shall locate an interval to the right of point *x* in each case in such a manner that it will not contain the point *b*.

Clearly, any such chain consists of a finite or countable set of intervals whose complement is at most a countable closed set.

The importance of chains of intervals is that one can, for example, (a) using these chains represent the increment of a continuous function $f(b) - f(a)$ in the form of a sum of increments over the intervals of the chain, and (b) express the variation of a continuous function as the

---

[10]An intuitive meaning of this assertion is as follows: if one can step from each point $x \in$ [*a, b*] to the right, then it is possible to move from *a* to *b* by no more than a countable number of steps.

limit of a sum of the absolute values of the increments over the intervals of the chain (see Leçons I, p. 54). These facts will be utilized later in the text.

Theorem 4.2 follows from the next, more general theorem proved by Lebesgue in Leçons I.

Theorem 4.2′

In order that a finite right upper derivative function $\bar{D}^+f$ of a continuous function f be summable, it is necessary and sufficient that f be a function of bounded variation. Moreover, Eqs. (4.19)–(4.22) are valid (see p. 72).

Proof [11]

We shall omit Lebesgue's proof of the measurability of upper derivatives of continuous functions. (A short proof is given in Leçons I, and a more detailed proof in Lebesgue [9]). Let

$$E_l \overset{\text{def}}{=} \underset{x}{E}\, (l \cdot \varepsilon < \bar{D}^+f \leqslant (l+1)\varepsilon),$$

where l is an arbitrary integer and $\varepsilon > 0$ (ε is fixed for the present). Choose positive numbers ε_l such that the series

$$\sum_{-\infty}^{+\infty} \varepsilon_l = \eta, \qquad \varepsilon \sum_{-\infty}^{+\infty} \varepsilon_l \cdot |l| = \delta$$

converge. Note that ε_l can be chosen so that η and δ are arbitrarily small. Consider for each l an open set G_l containing E_l and such that

$$mG_l - mE_l < \varepsilon_l.$$

Let n be a positive integer. For each l and each x, $x \in E_l$, define an interval of the form $(x, x + h)$ satisfying the following conditions:

$$l \cdot \varepsilon < \frac{f(x+h) - f(x)}{h} < (l+1)\varepsilon + \frac{\varepsilon}{n},$$

$$(x, x + h) \subset G_l, \qquad 0 < h < \frac{1}{n}. \tag{4.13}$$

---

[11] There are errors in the proof as given in Leçons I. The corrections are in Lebesgue [9]; see also Leçons II.

It follows directly from the definition of E_l that condition (4.13) can be satisfied for each $x \in E_l$ with h arbitrarily small. We now form a chain connecting a and b constructed from the above system of intervals. Denote by B_l the system of intervals in the chain, in which the left end points are the points E_l; clearly, $B_l \subset G_l$ and $mB_l \leqslant mG_l \leqslant mE_l + \varepsilon_l$. To bound mB_l from below we note that the measure of the part of the set E_l, which is contained in some $G_{l'}$, $l' \neq l$, is less than $\varepsilon_{l'}$; hence, the measure of the set of points in E_l, which fall into at least one $G_{l'}$, $l' \neq l$, is less than $\sum_l \varepsilon_l$, i.e., less than η. A point of the set $E_l - B_l$ belongs to some $B_{l'}$, $l' \neq l$ and hence, belongs to $G_{l'}$, $l' \neq l$ and therefore, $m(E_l - B_l) < \eta$. Thus,

$$\max[0, mE_l - \eta] \leqslant mB_l \leqslant mE_l + \varepsilon_l \qquad (4.14)$$

and

$$\sum_{l \geqslant 0}{}' (mE_l - \eta) l \cdot \varepsilon \leqslant \sum_{l=0}^{\infty} mB_l \cdot l \cdot \varepsilon \leqslant (\mathrm{I}) + \delta \qquad (4.15)$$

where the sum $\sum'$ denotes the sum of nonnegative terms and the sum (I) is defined by:

$$(\mathrm{I}) \stackrel{\text{def}}{=} \sum_0^{\infty} mE_l \cdot l \cdot \varepsilon$$

(the sum of the series here and below may equal $+\infty$). We also introduce the sum (II):

$$(\mathrm{II}) \stackrel{\text{def}}{=} \sum_{l=0}^{\infty} \left(\sum_{\Delta_i \in B_l} f(\Delta_i) \right).$$

For the increments $f(\Delta_i)$, $\Delta_i \subset B_l$, we have, according to (4.13),

$$m\Delta_i \cdot l \cdot \varepsilon \leqslant f(\Delta_i) < \left\{ (l + 1)\varepsilon + \frac{\varepsilon}{n} \right\} m \Delta_i. \qquad (4.16)$$

Hence,

$$\sum_{l=0}^{\infty} l \cdot \varepsilon \cdot mB_l \leqslant (\mathrm{II}) \leqslant \sum_{l=0}^{\infty} (l + 1)\varepsilon \cdot mB_l + \frac{\varepsilon}{n}(b - a). \qquad (4.17)$$

The sum (II) differs insignificantly from the sum (III) *of the non-negative increments* $f(\Delta)$ *over the intervals of the chain* Γ; indeed, the latter consists of the sum (II) and possibly of the sum of nonnegative terms $\sum'_{\Delta_i \subset B_{-1}} f(\Delta_i)$, which, in view of (4.14) and (4.16), is less than $(\varepsilon/n) (mE_{-1} + \varepsilon_{-1})$; therefore, as $n \to \infty$, the limits of the sums (II) and (III) are the same, and are equal to the positive variation of function f. We obtain from (4.15) and (4.17) that

$$\sum_l{}' (mE_l - \eta) \cdot \varepsilon \cdot l \leqslant (\text{II}) \leqslant (\text{I}) + \varepsilon(b - a) + \delta + \frac{\varepsilon}{n}(b - a);$$

as we send n to infinity the sum (II) converges to $V^+(f)$ and the extreme parts of the last inequality are independent of n, except for the term $(\varepsilon/n)(b - a)$. Hence we obtain

$$\sum_l{}'(mE_l - \eta) \cdot \varepsilon \cdot l \leqslant V^+(f) \leqslant (\text{I}) + \varepsilon(b - a) + \delta. \qquad (4.18)$$

Letting first η and δ and then ε approach zero, we finally obtain

$$V^+(f, [a, b]) = \int_{\substack{E(\bar{D}^+ \geqslant 0) \\ x}} \bar{D}^+ \, dx. \qquad (4.19)$$

Analogously, the inequality

$$V^-(f, [a, b]) = - \int_{\substack{E(\bar{D}^+ \leqslant 0) \\ x}} \bar{D}^+ \, dx \qquad (4.20)$$

is proved; hence,

$$V(f, [a, b]) = \int_a^b |\bar{D}^+ f| \, dx \qquad (4.21)$$

and

$$f(b) - f(a) = V^+ - V^- = \int_a^b \bar{D}^+ f \, dx. \qquad (4.22)$$

Theorem 4.2′ is thus proved.

Remark 1. With no change in the arguments, Eqs. (4.19)–(4.22) are proved for any of the three other derivative functions if they are finite everywhere.[12] Lebesgue observes that the reasoning which proves

---

[12] Equations (4.19)–(4.21) can be of the form $+\infty = +\infty$. In the case of bounded functions, the boundedness of each of the three variations implies the boundedness of the two others.

Theorem 4.2′ will also prove that *an arbitrary function of bounded variation has almost everywhere finite derivative numbers, which are summable on the set on which they are finite.*[13]

Remark 2. Lebesgue also utilizes Theorem 4.2′ to prove the following theorem. *The derivative of an indefinite Lebesgue integral exists and is equal almost everywhere to the integrand.*[14] In turn, this theorem is utilized to prove the existence almost everywhere of a tangent to a rectifiable curve, and hence, the differentiability almost everywhere of a function of bounded variation (Leçons, I Chapter VII).

4.8 CONCLUDING REMARKS

It would seem desirable in concluding to draw attention to certain general considerations due to Lebesgue in connection with his theory of integration. We shall, in particular, discuss some differences between the theory constructed in ILA (Lebesgue [2]) (1901–1902) and in Leçons I (1902–1903). This is probably of some interest if one wishes to trace the progress in Lebesgue's ideas during this period and to establish the role of previous results (especially Borel's) on the formation of Lebesgue's concepts.

4.8.1 Construction of a Measure

At that time Lebesgue considered his measure theory to be a completion of Borel's theory; this is especially noticeable in ILA. It is of interest to cite, for example, the following passage from the thesis: "... I define, following M. Borel, the measure of a set in terms of its basic properties. After achieving the completeness and rigor in the somewhat sketchy arguments given by M. Borel [reference to *Borel's* Leçons] I establish certain relations ..." (Lebesgue [2], p. 234).

On many occasions Lebesgue emphasizes the importance of B-sets as the only sets which can be arrived at in a constructive manner (Leçons I, p. 109 and the footnote). In ILA, Lebesgue notes the existence of a Borel kernel and a Borel envelope for each set and asserts that a

---

[13] The proof is based on Eq. (4.18); however in the preliminary arguments, the set $E_\infty \overset{\text{def}}{=} E_x(\bar{D}^+ = +\infty)$ should be taken into account.

[14] In particular, the well-known theorem on the points of density is proved.

measure can be assigned to L-measurable sets by the Borel method: he quotes Borel's Leçons, p. 48, which was given here in Chapter 3, pp. 46, 47.

4.8.2 Approach to the Integration Problem

An axiomatic approach to the construction of an integral is not yet available in ILA; in ILA, a construction (axiomatic) of a measure precedes the construction of the integral and hence, naturally, the geometrical definition of the integral precedes the analytical definition.

It seems that in Lebesgue's opinion the most interesting and desirable property of integrals is that which enables us to solve the fundamental problem of integral calculus: determination of a function from a known derivative (see ILA, Introduction). With apparent regret he notes that this problem for unbounded derivatives of functions cannot be solved in general using his integral. The prevalence of the idea of the primitive can be traced also in other sections of the thesis; in the first note in *Comptes Rendus* [1] this property also receives preeminence.

4.8.3 The Method of Chains

The method of chains in ILA is also absent; the proof of Theorem 4.2 presented there is based on the direct comparison of variational and integral sums; the awkwardness of this method apparently prohibited the establishment of a more general Theorem 4.2′ (and theorems mentioned in Remarks 1 and 2). We note that in due time the method of chains was replaced by Vitali's covering theorem and by other methods. (Although the chain method is not directly applicable in the multi-dimensional case, it is the simplest and most natural method in the linear case.)

4.8.4 Integration of Unbounded Functions

In ILA as well as in Leçons I, integrable functions were actually assumed to be finite; neither the geometrical nor the analytical definitions are applicable for functions which assume infinite values at certain points. This approach is insufficient, especially if one is concerned with the integration of derivative functions not finite everywhere. Therefore, Lebesgue is compelled to speak for example, not about the integrability of an almost everywhere finite derivative, but about its integrability on the set of points on which it is finite (Leçons I, p. 129). For the same reason, in the footnote on the last page of Leçons I, Lebesgue considers that property of functions which was called "absolute continuity" by Vitali, and suggests the following most general definition of the integral which includes the case of not-everywhere finite functions f.

Definition 4.2

A function f is called summable if there exists an absolutely continuous function F which possesses almost everywhere a derivative equal to f. Then, by definition, the integral of f from a to b is F(b) − F(a).

There is no discussion of this definition. Lebesgue asserts that it is applicable also to finite summable functions, i.e., *in order that a function F be an indefinite integral it is necessary and sufficient that it be absolutely continuous.* The first proof of this theorem is due to Vitali [2] and was published in 1905; it is also available in Beppo Levi's paper [1] (1906) and in Lebesgue's paper of 1907 [10]. In the latter, Lebesgue suggests that we define integrals of unbounded functions by the equality

$$\int_a^b f(x)\, dx \overset{\text{def}}{=} \lim_{N \to \infty} \int_a^b f \, {}_{-N}^{+N}(x)\, dx.$$

In the case of absolute convergence this definition is equivalent to Lebesgue's previous definitions.

4.8.5   Cardinalities of Measures

In ILA, Lebesgue presents an interesting comparison of (outer) Jordan and Lebesgue measures from the point of view of cardinalities of the sets required for their definition. For example, to define Jordan's outer measure for any set it is sufficient to have a monotonically decreasing sequence $\{\sigma_n\}$ of planar grids with sides $1/2^n$; having the given set E, one must choose for each n a system Δ_n of squares from σ_n containing points of E. The choice of Δ_n is determined by at most a countable number of trials (each trial determines whether or not points of E are contained in the given square of the subdivision σ_n), performed on the squares of the grid σ_n. Hence, the determination of a monotonic sequence $\{\Delta_n\}$ required for the definition of the outer Jordan measure is at most a countable process for any E.

On the contrary, from this point of view, the process of defining the Lebesgue outer measure of an a priori given set E is intrinsically un-countable. Indeed, by definition the outer Lebesgue measure is an infimum of a continuum of numbers—the measures of open sets G containing E.

Moreover, it may be noted that for any given countable collection $\mathscr{A}$ of open sets, there exists a set E, whose outer measure is not a limit of the measures of any sequence of open sets from $\mathscr{A}$ containing E.

Various versions of Lebesgue's definition of the integral are given in Chapter 6. Often definitions which do not use measure theory explicitly appear to be simpler. Here we should emphasize that no matter how the theory is presented one cannot escape the main difficulty—the consideration of a certain continual process. One can only make this difficulty seem less evident.

5 YOUNG'S INTEGRAL

5.1 YOUNG'S INTEGRAL

A very important contribution to the new theory of integration was made by an English mathematician, W. H. Young. His paper "Outline of the General Theory of Integration" [1], which was written about three years after Lebesgue's note in *Comptes Rendus*, contains a construction of the integral equivalent to Lebesque's construction. However, we must note at once that the reasons which prompted these two investigators to deal with generalizations of integrals were quite different and therefore the approach towards the solution of the problem is different. To clarify this point we quote the corresponding passages from Young's basic paper [1] ("Introductory," pp. 221–222): "The progress of the modern theory of sets of points . . . naturally leads us to put the question how far these definitions can be generalized. This theory has in fact taught us on the one hand that many of the theorems hitherto stated for finite numbers are true with or without modification for a countably infinite number, and on the other hand that closed sets of points possess many of the properties of intervals. We may, in accordance with these facts, divide the segment into an infinite number of nonoverlapping intervals . . .; or, more generally, into a finite or countably infinite number of sets of points.

"What would be the effect on the Riemann and Darboux definitions, if in those definitions the word 'finite' were replaced by 'countably infinite', and the word 'interval' by 'set of points'? A further question suggests itself: Are we at liberty to replace the segment (a, b) itself by a closed set of points, and so define integration with respect to any closed set of points?

"Going one step further, recognizing that the theory of the content of open sets[1] quite recently developed by M. Lebesgue[2] has enabled us to deal with all known open sets in much the same way as with closed sets as regards the very properties which come into consideration, we may attempt to replace both the segment and the intervals of the segment by any kinds of measurable sets. . . ."

Thus in his paper [1], Young sees two possible directions for generalizations: the first being the replacement of subdividing segments by general sets (we shall call them subdividing sets) and the second being an admission of countable subdivisions (and a combination of these two cases).

We now consider in more detail the manner in which Young implemented the program formulated in the quotation just given. (It is useful to compare Young's ideas with those of Peano and particularly of Jordan, see Chapter 2.) Functions appearing in the following discussion are all bounded.

First of all, Young shows that one cannot replace the subdividing segments in Riemann's definition by even the simplest sets (in a finite number): the limit as the measure of the subdividing set tends to zero (we emphasize *the measure* and not the diameter), generally may not exist even in the case of a continuous function. An example of this can easily be constructed. However, at the same time Young discovered that it is possible to generalize Riemann's definition using the second approach. He proved that *if a function f is R-integrable on* $[a, b]$ *then* $\int_a^b f(x)\, dx$

---

[1] Young uses the terms "open set" in the sense of "not necessarily a closed set," i.e., an arbitrary set.

[2] Young refers to the Lebesgue thesis [2] and to his own paper "Open Sets and the Theory of Content" [2], to be discussed further in Section 5.2.

is the limit of the sums $\sum_i f(\xi_i)\Delta_i$, where $\{\Delta_i\}$ is a countable system of mutually nonoverlapping segments and such that $m[(a, b) - \sum_i \Delta_i] = 0$. The limit is taken as $\max_i m\Delta_i \to 0$. The requirement that $m([a, b] - \sum_i \Delta_i) = 0$ is of course essential. (The assertion ceases to be true even if a summand of the type $f(\xi_i)m([a, b] - \sum_i \Delta_i)$ is included into the integral sum.) Young presents a corresponding example. This completes Young's investigations of Riemann's definition and he proceeds to a discussion of Darboux's definition.

(To aid in understanding Young's arguments in this connection we remind the reader that the upper and lower Darboux integrals are defined as the corresponding least upper and greatest lower bounds respectively, but at the same time it was shown by Darboux that these bounds are limits of the corresponding sums as the diameter of the subdivision tends to zero.)

First of all Young shows that: (1) The assertion analogous to the one given above for the Riemann integral also holds for the upper and lower Darboux integrals. (2) As in the case for the Riemann integral, the limit of the upper and lower sums does not exist if one allows subdivisions into arbitrary sets. In this connection, Young investigates two versions of the definition of Darboux upper and lower integrals. These are as follows.

Definition 5.1

Let $\{E_i\}$ be a finite or countable subdivision of the segment $[a, b]$ into measurable sets, $[a, b] = E_i$, $E_i \cdot E_j = 0$, $i \neq j$; let $M_i = \sup_{x \in E_i} f(x)$, $m_i = \inf_{x \in E_i} f(x)$. Then the numbers $\inf \sum_i M_i mE_i$, $\sup \sum_i m_i mE_i$, where the supremum and infimum taken over all possible subdivisions $\{E_i\}$ are called the upper and the lower integrals, respectively, of the function f.

Young first assumes that this definition will be justified if one can prove that (1) the upper integral is not less than the lower one, and (2) these numbers agree in all cases with the extreme Darboux integrals. He verifies that (1) holds,[3] but at the same time a simple example shows

---

[3] This verification is standard: it is shown that (a) the upper sum is not less than the lower sum for the same subdivision, (b) as subdivisions become more and more refined the sums vary monotonically, and finally, (c) each upper sum is not less than any one of the lower sums from which assertion (1) follows immediately.

that (2) is not generally valid. However, one important case can be singled out, the case when f is integrable. Here (2) is fulfilled and the extreme integrals in the sense of Definition 5.1 agree with the Riemann integral. The following case when (2) is fulfilled is studied in detail by Young, the case when f is semicontinuous. He shows that if f is upper (lower) semicontinuous, then its upper (lower) Darboux integral agrees with the upper (lower) integral in the sense of Definition 5.1. (In what follows, we shall consider only the case of upper semicontinuous functions and upper integrals; all the assertions can be reformulated for lower semicontinuous functions and lower integrals with the obvious modifications.)

We now note that the upper Darboux integral of the function f agrees with the upper integral of the function $\varphi(x) = \overline{\lim}_{x' \to x} f(x')$. But φ is an upper semicontinuous function! This immediately indicates what must be done in order that (2) (i.e., the required equivalence of the definitions) holds: in Definition 5.1 one should form the upper sum not for f but for φ. This is the approach taken by Young. The following is the new definition proposed by Young:

Definition 5.2

Let S be a set of a positive measure, and let $\{E_i\}$ be a subdivision of S into measurable sets. Let $\varphi(x) = \overline{\lim}_{x' \to x} f(x')$, $M_i = \sup_{x \in E_i} \varphi(x)$. Then the number $\inf \sum M_i mE_i$, where the infimum is taken over all possible systems $\{E_i\}$, is called the upper integral of the function $f(x)$ on the set S.

It is clear from Definition 5.2 that in the case when S is a segment, the extreme integrals in the sense of the definition agree with the extreme Darboux integrals. This agreement also holds in the general case if the upper Darboux integral of f on the set S is to be interpreted as the infimum of the upper sums $\sum M_i m(S \cdot \Delta_i)$ over all possible systems of mutually nonoverlapping segments $\{\Delta_i\}$ such that $S \subset \sum_i \Delta_i$; moreover, here $M_i = \sup_{S \cdot \Delta_i} f(x)$. Young shows that upper sums of this type converge to the corresponding upper integral; namely, for every $\varepsilon > 0$, a δ can be found such that if $m(S - \sum_i \Delta_i) + m(\sum_i \Delta_i - S) < \delta$ and $m \Delta_i < \delta$ for any i, then $\sum_i M_i m(S \cdot \Delta_i) - I < \varepsilon$, where $I = \inf \sum_i M_i m(S \cdot \Delta_i)$.

Definition 5.3

When the upper and lower integrals in the sense of Definition 5.2 agree, the function is called integrable on the set S and the common value of these integrals is called the integral of the function on the set S.

Young's goal is thus achieved: he has found a definition of extreme Darboux integrals in a form which allows generalization in the two directions stated at the beginning of this section.

Concluding his investigations connected with Darboux and Riemann's definitions, Young constructs important formulas which express the extreme integrals in terms of Riemann's integrals for the case of bounded functions. They are as follows.

Let S be a set, $k \leqslant f(x) \leqslant k'$, $\varphi(x) = \overline{\lim}_{x' \to x} f(x')$, and let $G_{-1} \overset{\text{def}}{=} 0$,

$$G_r \overset{\text{def}}{=} \underset{x \in S}{E} \left(\varphi(x) \geqslant k' - \frac{k' - k}{n} r \right), \qquad r = 0, \ldots, n;$$

then

$$\sum_{r=0}^{n} m(G_r - G_{r-1})\left(k' - \frac{k' - k}{n}(r - 1) \right) = \sum_{r=0}^{n} \frac{k' - k}{n} mG_r + kmG_n$$

is an upper sum and hence its limit as $n \to \infty$ is not less than the upper integral; this limit is equal to the quantity

$$\int_k^{k'} I(k) \, dk + k \cdot mS, \qquad (5.1)$$

where $I(k) \overset{\text{def}}{=} m E_{x \in S}(\varphi(x) \geqslant k)$. On the other hand, any upper sum is not less than the lower sum

$$\sum_{r=0}^{n} m(G_r - G_{r-1})\left[k' - \frac{k' - k}{n} r \right],$$

whose limit is also the quantity (5.1). It follows from here that this expression equals exactly the upper integral. As a corollary we obtain that the lower integral is equal to

$$k'mS - \int_k^{k'} J(k) \, dk, \qquad (5.2)$$

where $J(k) = m E_{x \in S}(\varphi(x) \leqslant k)$. [The functions $I(k)$, $J(k)$ are monotonic and hence are R-integrable.]

After a detailed study of the most general forms of definitions of the
Riemann–Darboux integral, Young returns to Definition 5.1. Here he
notes that although this definition does not agree with the usual definitions, it is nevertheless more natural than Definition 5.2; he therefore
suggests "to throw overboard" the definitions given by Darboux and
Riemann and to define integration on the basis of Definition 5.1. The
corresponding extreme integrals are called *the upper and lower Young
integrals* and if they agree, their common value is called *the Young
integral* of the function f on the measurable set S.

The agreement between the Lebesgue and the Young integrals for
the case of a measurable function f is proved by Young from expressions
(5.1) and (5.2). In fact, the proof which was used in their derivation is
applicable without change if instead of the function φ a bounded
measurable function f is considered. This shows that the extreme Young
integrals agree with (5.1) [or (5.2)], which is, by its construction, a
Lebesgue integral (the limit of Lebesgue integral sums).

Independently of this argument, Young directly establishes the
equivalence of his definition with Lebesgue's geometrical definition;
hence *only measurable functions are integrable in the Young sense.*

Young emphasizes that expressions (5.1) and (5.2) can serve as
definitions of Lebesque integrals; if however there are difficulties in
the construction of functions $I(k)$ and $J(k)$, then he prefers his definition
in terms of extreme sums. Moreover, he considers as an advantage of
his definition the fact that S can be an arbitrary not necessarily bounded
measurable set. Lebesgue's continuation of a function from the set S
to a segment containing it by defining the function as zero outside S is
considered somewhat inadequate by Young. This is because certain
properties of functions which simplify integration such as continuity
and semicontinuity are lost as a result of this continuation.

5.2 YOUNG'S MEASURE THEORY

The role and significance of Young's contributions toward the
construction of a new integration theory cannot be properly evaluated
without discussing his paper "Open sets and the Theory of Content,"
published in 1904 (Young [2]), before the paper we discussed in the
previous section (Young [1]). In this work, Young constructs a class of

sets which agrees with the class of L-measurable sets, but has no connection with the problems of integration (at least there is no mention of this connection in the paper).

The following theorem probably served as a motive for the paper: *if an infinite sequence of systems of intervals (open sets) each of total measure $\geq \sigma$ is given on $[a, b]$, then the upper limit of this sequence[4] contains a closed set of measure $\geq \sigma - \varepsilon$, where $\varepsilon > 0$ is arbitrary* (Borel [3] discussed this theorem just a year before that, see Section 6.2). Young proves the theorem and extends it to the case of a sequence of closed sets of measure[5] $\geq \sigma$ and finally in the most general form to the case of a sequence of arbitrary sets. In this manner, Young arrives at the definition of the inner content $m_i E$ of the set E. He notes immediately that this notion was first introduced by Lebesgue. Young is concerned as to what extent this measure preserves the additivity property

$$m_i E + m_i M = m_i (E + M). \tag{5.3}$$

He distinguishes the class of sets E which satisfy relation (5.3) for any set M, $M \cdot E = 0$ and then proves that (a) the (finite) arithmetic operations performed on the sets of this class produce sets of the same class and, of particular importance, (b) this property is preserved by taking limits of monotonic sequences of sets (i.e., if $\{E_n\}$ is a monotonic sequence of sets in a class, then $\lim E_n$ also belongs to this class).

It follows from (a) and (b) that the class of sets defined by Young has countable additivity and oddly this property was not explicitly noted by Young. However, Young observes that the additivity property of a class [which Young interprets as properties (a) and (b)] permits us to introduce a classification of sets; this idea was developed in his later paper [3].

Analogously a class of sets E satisfying the relation

$$m_0 E + m_0 M = m_0 (E + M) \tag{5.4}$$

is investigated; here $m_0 E$ is the outer content defined by Young in the same manner as Lebesgue. Thus an inner additive class, an outer additive class, and their common part, the additive class are constructed. Young proves the Lebesgue measurability of bounded sets of the class.

---

[4] The upper limit (or limit superior) of a sequence of sets is a set consisting of points each of which belongs to infinitely many sets of the sequence.

[5] Young's definition of the measure of closed sets agrees with Cantor's definition.

He notes that the following question is open: does his additive class contain all the sets, or rather do nonmeasurable sets exist?

Concluding our brief description of Young's measure theory we present his remarks on Lebesgue's measure theory:

(1)   Lebesgue's results become known to Young after his paper was written.

(2)   Young considered Lebesgue's theory somewhat more limited than his own, since only bounded sets were used; moreover, he believed that the definition of measurability using relations (5.3) and (5.4) better revealed the nature of measurable sets (in their relation to arbitrary sets M).

5.3   THE INTERRELATION BETWEEN LEBESGUE'S AND YOUNG'S CONTRIBUTIONS

We have studied the progress of integration theory during the period 1901–1904 because the value of these accomplishments in the theory of functions and in modern mathematics in general cannot be overestimated. These achievements are associated with the names of Lebesgue and Young. (We should also mention Vitali [3] who in the period 1903–1904 arrived independently at a definition of measure identical to Lebesgue's definition and proved the basic properties of the measure. His investigations, however, were not connected with integration, and were aimed at generalizing the notion of length for arbitrary sets by means of a process which is more general than Jordan's or Borel's.) Based on the discussions given in the previous sections, we have grounds for comparison of the relative merits of Lebesgue's investigations versus Young's investigations.

The main achievement of Lebesgue is not the fact that he extended the notion of length (area) to a wider class of objects and that it was done several years before Young's papers appeared. The basic result in his theory is that the Lebesgue measure is the unique solution of the measure problem for the class of L-measurable sets. The same can be said about Lebesgue's integral. Here is the corresponding quotation from Leçons II, p. 100, footnote 1: "The principal advantages of the arguments about measurable sets . . . are not that a wider class of sets is considered but that they originate from the basic property of sets to which a measure can be ascribed. . . ."

Furthermore, Lebesgue did not limit his investigation to the construction of a theory but also studied its applicability to classical problems associated with integration such as determination of primitive functions and areas of planar sets (including also the curve lengths).

As we have seen above, no such general approach is present in Young's definitions of the measure and the integral. His aim was to generalize the Riemann and Darboux integration procedures as much as possible. He arrived at the same result as Lebesgue, but what is the significance of this discovery? The possibility of term-by-term integration of sequences, which is such an important property of integrals (connected with the countable additivity of the measure) is absent in Young's work. There is also no mention of applications of integrals (at least in those papers which appeared before 1905; however, by then he had become acquainted with Lebesgue's thesis).

Thus, although the final results of Young's and Lebesgue's investigations are the same, the problems which were solved by Lebesgue in the course of his investigations are fundamentally broader and deeper.

The advantage of Young's definition (as emphasized by the author) that unbounded sets are also allowed is immaterial: the transition to unbounded sets in the theory of measure and hence also in the theory of integrals is straightforward, as the following quotation from Lebesgue ([13], p. 192) indicates: ". . . the theory which attacks fundamental difficulties, and which is not concerned with the accompanying arguments, is created in the first place . . ."

Also, Lebesgue's continuation of a function from a set to the whole axis or to an interval has become generally acceptable, and Young's criticism in this connection is not justified.

At the same time, we emphasize that there are several significant merits in Young's theory. His definition of an additive class by Eq. (5.3) or (5.4) became a useful tool for the subsequent development of measure theory (the general form of this definition was utilized in Carathéodory's theory). Other advantages, such as direct analogy to the Darboux definition and greater freedom in the choice of integral sums, contributed to making Young's definition acceptable, along with Lebesgue's definition. [Another valuable feature of Young's investigation is that it shows that Lebesgue's definition, which at first glance seems completely original and "unorthodox," can actually be obtained from the classical definition by a very natural generalization.]

6 OTHER DEFINITIONS RELATED TO THE DEFINITION OF LEBESGUE'S INTEGRAL

In the preceding chapters we described the period during which the foundations of modern theory of functions of a real variable were laid, the decisive role played by Borel's ideas, and the contributions of Lebesgue and Young. The succeeding period was the era during which these ideas and methods were assimilated. This was accomplished by the end of the second decade of the twentieth century. In what follows, we shall describe investigations in which the Lebesgue definition of the integral was further studied and modified. These modifications are due to Young, Borel, Riesz, Pierpont, Denjoy, and others. They emerged first in order to bring the Lebesgue definition of the integral closer to the more conventional Riemann (or Darboux) definitions, and second, to make the exposition more elementary, in other words, to make it closer to the concepts of classical analysis.

6.1 YOUNG'S FIRST DEFINITION

We are already familiar with the ideas that guided Young in writing his "Outline of the General Theory of Integration" (Young [1]): the new integral was obtained as a modification of the classical Darboux integral. [We may add to the discussion given in Chapter 5 that while the form of the integral in the Darboux sense has been retained, Young's definition is more complex not only because countable subdivisions are allowed but mainly because in place of the segments, the new elements of subdivision, the Lebesgue-measurable sets, have much more complex configurations. Roughly the same remarks can be made about the integral defined by Eqs. (5.1) and (5.2), p. 81.

6.2 BOREL'S INTEGRAL

We shall discuss Borel's ideas concerning integration published in 1910 in two notes in *Comptes Rendus* (Borel [5], [6]). In [5] Borel remarks that his writing of these notes was prompted by the reading of Lebesgue's paper [15] (we shall discuss this paper in Section 6.8); the aim of this note is to bridge the gap between the Lebesgue and the Riemann definitions. We deliberately used the term "ideas" since we do not regard the content of these notes as a precise formulation of an integral, but rather as the rudiment of a definition which generalizes the Harnack–Jordan definition.[1] Harnack's definition is based on two main features: (1) a set T of singularities is given (in Harnack's terminology, the set of points of unboundedness) which depends on the function; (2) finite coverings of the set T by intervals are considered; if we want to adapt this definition, then for any system of covering intervals, the function must be R-integrable on their complementary segments. It is clear from Borel's considerations that the second point should be modified: the finite covering by intervals should be replaced by countable coverings. However, Borel does not indicate with sufficient clarity which set of singularities he has in mind. Here is the quotation from Borel's paper: "Consider a function which has infinitely many discontinuities between a and b and assume that it is possible to include

---

[1] Borel does not associate his definition with the Harnack–Jordan definition.

them in an infinite system of intervals (α_n, β_n) in such a manner that the previous definition will be applicable . . ." The end of the last sentence means as follows: the Riemann sum $\sum f(\xi_i)h_i$ is constructed by a subdivision $\{x_i\}$ in such a manner that the points of the subdivision x_i and the points $\xi_i \in [x_{i-1}, x_i]$ do not belong to the intervals (α_n, β_n); h_i is the *reduced* length of $[x_{i-1}, x_i]$, equal to the difference between the length $[x_{i-1}, x_i]$ and the length of those intervals (α_n, β_n) which are contained in $[x_{i-1}, x_i]$. We quote further: "If, for any 'system of excluded intervals' (*intervalles d'exclusion*, in the original French text), the Riemann sums (constructed as stated above) tend to a limit as the lengths of the intervals $[x_{i-1}, x_i]$ tend to zero,[2] then the limit of these limits as the sum of the excluded intervals tends to zero is precisely an integral in the Lebesgue sense at least for bounded functions. For unbounded functions our definition is more general than M. Lebesgue's definition . . ." (Borel [5], p. 201).

Borel concedes that the proposed method is not as natural as that of Lebesgue; however, he believes that in certain specific cases, as for example in the case when the function is given in the form of a sum of a series of functions with poles, his definition is more direct.

Two questions arise in connection with the definition as given by Borel [5]: the first, which points of discontinuity are considered and the second, how is the choice of the "intervalles d'exclusion" made? The second note by Borel [6] was intended to clarify the contents of his definition. Moreover, Borel aimed in this note to show how the idea of defining the integral could be deduced from his earlier investigations without reference to Lebesgue's works. He recalls the definition of measure given in his paper [1] (see Chapter 3) to which the following condition is now specifically added: if an arbitrary set is contained in a measurable set, then its measure is less than or equal to the measure of the measurable set (see Chapter 3, p. 45). From this he concludes that if some set can be included into a system of intervals of an arbitrarily small total length, then it is of measure zero. Furthermore, Borel refers to a theorem formulated in his paper [3] of 1903 and proved in his [4] "Leçons sur les fonctions des variables réelles," Paris, 1905. (He

---

[2] The reduced lengths h_i are meant here.

intended to justify the consistency of his definition of measure by this theorem.) The theorem is as follows.

Theorem 6.1

Let E_n be a countable sequence of measurable sets, situated in a finite interval. Then $\overline{\lim}_{n \to \infty} mE_n \leqslant m(\overline{\lim}_{n \to \infty} E_n)$.

The importance of this theorem was noted by several authors. It was proved by Young [2] in 1904 (see Chapter 5, p. 83); he also remarks on the origin of this theorem.

As a corollary to Theorem 6.1, Borel formulates (again without proof) the following theorem and its corollary:

Theorem 6.2

The property of a function to be continuous, except on a set of arbitrarily small measure, is preserved in the limit.[3]

Corollary

All analytically representable functions, (i.e., functions of the Baire classification) possess this property.

Borel, referring to Theorem 6.2, notes that every bounded analytically representable function is integrable in the sense of his definition (since it is continuous outside of open sets of an arbitrarily small measure). Thus the conclusion is clear: the "intervalles d'exclusion" are such that the function is continuous (or Riemann-integrable) outside of them. If these were all the conditions imposed on the "intervalles d'exclusion," the integration in the Borel sense would have been equivalent to integration in the Lebesgue sense (to show this it is sufficient to utilize the C-property of measurable functions and the absolute continuity of Lebesgue's integral) for bounded as well as for summable

---

[3] Here is the exact formulation of this property: for every $\varepsilon > 0$ there exists a measurable subset $e \subset [a, b]$, $me < \varepsilon$, such that the function is continuous on $[a, b] - e$. Lusin proposed to call this property of a function *the C-property*. This C-property of functions measurable in the Lebesgue sense was known to Lebesgue. It was fully proved by Vitali [1] in 1905.

functions. However, when reading this note by Borel, one gets the impression that he admitted a particular choice of the "intervalles d'exclusion" in certain cases depending on the individual characteristics of the function.

When analyzing Borel's definition, we must note the condition expressed by Borel in his first note in *Comptes Rendus* which was imposed on the "intervalles d'exclusion": that they must contain points of the singular set. Of course one cannot take literally the term "points of discontinuity" used by Borel, since in this case the functions under consideration would have been continuous almost everywhere. Undoubtedly, Borel had in mind not only these functions. The idea of the existence of some singular set was clearly expressed in Borel's following paper [7] in 1912 which is discussed in Section 6.3. At present, we note that an indication of the feasibility of defining an integral according to the procedure given above (without mentioning the set of singularities[4]), as a limit of integrals of functions integrable in the Riemann sense, was given by Lebesgue in his earlier note [7] in 1903. This article was written in connection with Borel's note and Theorem 6.2 discussed above. Here Lebesgue asserts that Theorem 6.2 follows from another theorem proved by him[5] on the property of a measurable function being approximatively continuous almost everywhere and that this property is equivalent to the condition of measurability of the function. Furthermore, it is stated that if Theorem 6.2 is given in the form: every measurable function f differs from an integrable function φ_ε on a set of measure less than ε, and taking the validity of term-by-term integration of uniformly bounded sequences into account, one can then define an integral of bounded measurable functions by the integrals of functions φ_ε.

6.3 CONTINUATION

In 1912 in the *Journal des Mathématiques*, Borel [7] gave the following modified definition of the integral. We present it here verbatim.

---

[4] From the content of Borel's note [6] it follows that he had in mind only the set of systems of "intervalles d'exclusion"; he does not mention a singular set which should be contained in each system.

[5] This theorem does not appear in Leçons I; it was given in the course presented by Lebesgue in the Collège de France.

Definition 6.1

Let f be an unbounded function nonintegrable in the Riemann sense; assume that a countable set of points $A_1, A_2, \ldots, A_n, \ldots,$ can be defined in the domain of integration with the following property: if the intervals (B_n, C_n) are such that $A_n \in (B_n, C_n)$ and the series $\sum m(B_n, C_n)$ is convergent with its sum equal to ε, then the generalized Riemann sums tend to a limit for any of these intervals, and these limits in turn tend to a limit as ε tends to zero. The latter integral is by definition the Riemann integral in the generalized sense.

The Riemann sums are those of the form

$$\sum h_i f(\xi_i),$$

and it is assumed that:

(1) The points of subdivision x_i do not belong to the intervals of exclusion (B_n, C_n);

(2) h_i is equal to the number $x_i - x_{i-1}$ minus the total length of the excluded intervals contained in (x_{i-1}, x_i);

(3) ξ_i a point between x_{i-1} and x_i but does not belong to the excluded intervals.

Following this definition, Borel asserts that the integral as defined by Definition 6.1 agrees in the multidimensional case with the Lebesgue integral and in the one-dimensional case is more general than the latter (in view of the cases of conditional convergence). The definition given above, in contrast with the previous one (described in the beginning of Section 6.3) is in its final form; no attempts were made by Borel at that time to consider possible variations of this definition.

*We shall now analyze this definition and its relation to the Lebesgue integral. We first note that values of the function taken on the singular set $\{A_n\} = N$ have no bearing either on the integrability or on the value of the integral. Therefore, the integrability of the function is determined by its behavior on $[a, b] - N$.

Assertion 1

Every bounded function to which Definition 6.1 is applicable is Lebesgue-measurable.

Proof

Such a function coincides with an integrable function and hence with an everywhere measurable function, except possibly on a set of arbitrarily small measure.

Assertion 2

Let f be a bounded function on [a, b] and let N be a measurable set. In order that f be integrable in the Riemann sense outside of every open neighborhood G of the set N, it is necessary and sufficient that f be continuous almost everywhere on the set [a, b] − N.

Assertion 3

The class of functions bounded and integrable in the sense of Definition 6.1 coincides with the class of functions which are bounded and which differ from R-integrable functions on a countable set.

We now consider the structure of unbounded functions integrable in the sense of Definition 6.1.

Assertion 4

In order that f be bounded outside every open neighborhood G of the set N, it is necessary and sufficient that f, considered on [a, b] − N, have no points of unboundedness on this set.

Thus, the set $\tilde{E}^{\infty}$ of points of unboundedness of a function integrable in the sense of Definition 6.1 (we ignore the values of this function on the set $\{A_i\}$ mentioned in the definition) is contained in $\{A_i\}$ and is therefore countable; the structure of f inside the intervals adjacent to $\tilde{E}^{\infty}$ is described by Assertion 3. Finally, we obtain Assertion 5.

Assertion 5

Every function integrable in the sense of Definition 6.1 is integrable in the Harnack sense after its values are modified on a countable set.

Clearly, there are H-integrable functions which cannot be converted into functions integrable in the sense of Definition 6.1 by changing their values on a countable set.*

Assume now that (in the spirit of the notes in *Comptes Rendus*) only those "intervalles d'exclusion" are taken into account outside of which the function is Riemann-integrable. This modification of Definition 6.1 we call Definition 6.2. Then every summable function is integrable in the sense of Definition 6.2. Indeed, if in place of $\{A_i\}$ we consider an everywhere dense set on $[a, b]$, then there is actually no restriction on the choice of intervals (B_n, C_n) and it is sufficient to repeat the arguments presented on p. 89.

We also note that the condition of countability of the singular set N appearing in Definition 6.2 is not important, since one can always choose a countable everywhere dense set on N for the singular set. It is important only that $mN = 0$.

Evidently, certain nonsummable functions are integrable in the Borel sense, particularly in the case of the conditional convergence of the integral, as well as all the functions conditionally integrable in the Harnack sense, for example. These functions are integrable in the sense of Definition 6.2. We shall encounter them in Part III of this book where Denjoy integrals are discussed.

6.4 ADDITIONAL REMARKS ON BOREL'S DEFINITIONS

Primarily because of their ambiguity, Borel's definitions were investigated in several papers. Hahn [1] (1915) assumed that there exists a set of singularities N of measure zero, and considered the exclusion intervals outside of which R-integrability holds. As we have seen above, these assumptions correspond to Definition 6.2. In this connection the following interesting theorem is proved.

If f is integrable in the Borel sense with an a priori given set of singularities N, then on the closure of N, the function f is necessarily integrable in the Lebesgue sense, and the series $\sum_v \omega_v$ is convergent, where

$$\omega_v \overset{\text{def}}{=} \sup_{(\alpha, \beta) \subset (a_v, b_v)} \left| \int_\alpha^\beta f(x)\, dx \right|$$

and $C\overline{N} = \sum (a_v, b_v)$; moreover, the following formula

$$(\text{B}) \int_a^b f\, dx = \sum \int_{a_v}^{b_v} f\, dx + (\text{L}) \int_N f\, dx \qquad (6.1)$$

is valid.

[This formula corresponds to Moore's formula for Harnack's integral, see Eq. (2.5), p. 21.] Furthermore, Hahn shows that the Borel integral (B) $\int_a^b f(x)\, dx$ is independent of the choice of the singular set N (provided only that the integral exists for such a choice—the integral may exist for one choice of N and not for another). Thus the theorem proved by Hahn characterizes completely functions integrable in the sense of Definition 6.2.

Borel's integral was the subject of detailed investigations in Lusin's [4] thesis "The Integral and Trigonometric Series" (1916). Lusin considers the definition from the French edition of *Encyclopédie des Sciences Mathématiques* edited by Montel, which agrees with Borel's definition in his first note in *Comptes Rendus* [5] without requiring the existence of a singular set. Furthermore, he investigates Definitions 6.1 and 6.2 and shows that every function integrable in the sense of Definition 6.1 is integrable in the Dirichlet–Lebesgue sense. (Moreover, it is integrable in the Harnack–Lebesgue sense; see the corresponding definitions in Chapter 8.)

We have seen during our discussion of Harnack's integral that there exists a function integrable in the Dirichlet sense for which the limit (2.3) (see page 18) does not exist; therefore, the limit in Definition 6.1 certainly does not exist. (This is easily verified if one notes that E^∞ necessarily belongs to the closure of the sequence $A_1, A_2, \ldots, A_n, \ldots$, given in Definition 6.1.) An example of this type can be found in Men'shov [1]. Thus there exists a function integrable in the Dirichlet sense but not integrable according to Borel in the sense of Definition 6.1.

Borel's assertion that his integral is more general compared to Lebesgue's integral was the cause of the dispute between Borel and Lebesgue in the pages of *Annales de l'Ecole Normale Supérieure* **35** (1918), **36** (1919), **37** (1920) (Borel [8], [9] and Lebesgue [13], [14]). Several assertions and examples presented by Lebesgue (of the type of Assertions 1–5) were aimed at refuting Borel's assertion.

In the final result this controversy concerning the definition of the integral became a dispute concerning priority in the construction of measure theory (and hence of the integral). In the course of these arguments both participants expressed their views on a number of more general problems; in particular, opinions were expressed about the type of infinity which may be used in the construction of mathematical theories and so on. It is not our purpose to analyze the general views of Borel and Lebesgue on these topics; we would like only to stress those points which are important for understanding the Borel integral: when constructing a theory, Borel considered it obligatory to remain within the bounds of at most countable processes and he preferred not to consider such a process as a whole (the actual infinity!) but in its formation. The construction of sets measurable in the Borel sense from the intervals by means of a countable number of additions and subtractions may serve as a classical example. Borel's desire to determine some countable process which would enable us to obtain a general integral from the Riemann integral is therefore understandable.

Thus the definition of the integral given by Borel [7] (Definition 6.1)[6] should be considered the final form of his definition. The type of definition suggested by Hahn cannot be considered an extension or modification of Borel's definition: while preserving its form, it basically alters the content of Borel's concept.

In the *Journal de Mathématiques*, Borel [7] suggests a definition of the integral for the case of bounded functions based on the notion of a function asymptotically equivalent to a polynomial. According to Borel, the function is *asymptotically equivalent to a polynomial* if for every $\varepsilon > 0$ and $\delta > 0$ there exists a polynomial $P(x, \varepsilon, \delta)$ in x such that $mE_x(|P(x, \varepsilon, \delta) - f(x)| > \varepsilon) < \delta$. Correspondingly, a sequence P_n is called *asymptotically convergent* to f if

$$\lim_{\substack{n \to \infty \\ x}} mE(|P_n(x) - f(x)| > \varepsilon) = 0$$

for every $\varepsilon > 0$.

Borel proposes to define an integral of a bounded function f by the equality

$$\int_a^b f(x)\, dx \overset{\text{def}}{=} \lim_{n \to \infty} \int_a^b P_n(x)\, dx, \tag{6.2}$$

where P is a uniformly bounded sequence of polynomials asymptotically convergent to f.

---

[6] Hildebrandt [1], as well as Hahn, considers the singular set to be a set of measure zero.

Borel proves the theorem that the property of a function to be asymptotically equivalent to a polynomial is preserved in the limit (this theorem was in essence formulated in his note [3] in 1903, see above) and concludes that the definition, using Eq. (6.2), is applicable to all bounded analytically representable functions.

Recall Weierstrass' theorem that every continuous function on $[a, b]$ is the limit of a uniformly convergent sequence of polynomials; this theorem shows in particular that in Borel's definition the polynomials may be replaced by continuous functions, and this is the form in which the definition was stated by Lebesgue [7] in 1903.

6.5 RIESZ' DEFINITION

Riesz' note [1] was published in 1912 in *Comptes Rendus*; it was preceded by Borel's note [10] appearing in the same volume. In his note Borel suggests that it is desirable to simplify the exposition of measure and integration theory to make it suitable for inclusion in any textbook on mathematical analysis. He believed that this aim could be achieved by using his definitions.

Riesz, however, remarks that it is difficult to decide which theory should be considered elementary. A criterion for this decision may be the ease with which the theory is adapted. Riesz considered that the necessity of preliminary study of measure theory was the main obstacle for comprehension of Lebesgue's integral. In view of this fact he proposed an integration theory which utilized the notion of measure as little as possible. In Riesz' definition only the notion of a set of measure zero is used (see Remark, Chapter 4, p. 72). Here we present the outline of his theory.

A *simple function* is a function with a finite number of discontinuities on $[a, b]$ which assume a finite number of values $C_1, C_2, \ldots, C_n$. The integral of such a function is the Riemann integral. A *summable function* (cf. Section 4.5) is the limit of a uniformly bounded sequence of simple functions convergent almost everywhere.

Reisz' Definition

Let a (summable) function f be a limit almost everywhere of a uniformly bounded sequence of simple functions $\{f_n\}$. *Then the integral of* f *is the limit of the integral of functions* f_n *as* $n \to \infty$.

In order that this concept be well defined, it is necesssary to show (without using measure theory) that for an almost everywhere convergent and uniformly bounded sequence of simple functions, the integrals converge to a limit which depends only on the limiting function.

The case of integrability in the Riemann sense is treated separately: *in order that a summable function* f *be integrable in the Riemann sense it is necessary and sufficient that there exist a sequence of simple functions which converges to* f *uniformly at almost all points.*[7]

*A few remarks are in order about Riesz' definition.

From the properties of measurable functions and Lebesgue's theorem on term-by-term integrations it follows that the functions summable in Riesz' sense are bounded L-measurable functions, and also that Riesz' integral has the same value as the Lebesgue integral. To establish the complete equivalence of these two definitions it is sufficient to show that every measurable bounded function is a limit almost everywhere of a sequence of simple functions. But this follows from the fact that every R-integrable function is such a function (see Section 6.6 for more details) and that, as has already been noted on several occasions, every bounded L-integrable function is a limit amost everywhere of a sequence of R-integrable functions.

As far as integrability in the Riemann sense is concerned it is sufficient to observe that if a sequence $\{f_n\}$ of functions continuous at x_0 converges uniformly to f at the point x_0, then f is continuous at x_0. Since a sequence of step functions is simultaneously continuous everywhere, except possibly at the points of a countable set, the limit function f is bounded and is almost everywhere continuous, i.e., is integrable in the Riemann sense.

On the other hand, every R-integrable function f is a limit almost everywhere of a monotonically decreasing sequence of step functions uniformly convergent at the continuity points of f.*

[7] A sequence $\{f_n\}$ is called uniformly convergent at point x_0 to f if, for every $\varepsilon > 0$, an N and a neighborhood Δ of the point x_0 can be found such that for all points $x \in \Delta$ the inequality $|f_n(x) - f(x)| < \varepsilon$, $n > N$ is valid.

6.6   YOUNG'S SECOND DEFINITION

In 1910, Young [4] proposed a definition of an integral based on the *method of monotonic sequences*. Certain methodological improvements [5] allowed him to exclude the notion of the measure of a set from his theory. It is noteworthy that the second paper was written, as Young puts it, at the request of "a distinguished mathematician of the older school": in those days the Lebesgue theory and the new method were not as yet widely accepted.

We shall now discuss the Young method of monotonic sequences. A careful study of integration theory of continuous functions led Young to the conclusion that this theory is based on two principles: (1) the integral is a limit of integrals of step functions and these step functions can be taken to be upper(lower) semicontinuous if the value of each of these functions at a discontinuity point is attached to the higher(lower) step; (2) the limit function is approximated by means of monotonic sequences of step functions. (Indeed, for example, the upper Darboux sum $\sum M_i \Delta x_i$ is exactly the integral of a step function equal to M_i on the set Δ_i!) Next, Young points out that with such an interpretation of the integration process one should assume that the integration (the lower and the upper) of a given function is replaced by the integration of two semicontinuous functions; this concept also appears in Young [1] [see Chapter 5, the functions $\overline{\lim} f(x)$ and $\underline{\lim} f(x)$, p. 80].

These principles resulted in the integration process described here in three steps:

(1)   The simple u- and l-functions, i.e., upper semicontinuous and correspondingly lower semicontinuous step function (using Young's notation here and in the following), are taken as the initial elements for the construction of the integral. The integrals of these functions are by definition Riemann integrals. Before proceeding further one must show that every monotonic sequence of simple functions converging to a simple function can be integrated term by term and also that the sums and inequalities formed from the step functions can be integrated term by term.

(2) u-Functions(l-functions) are the limits of monotonically decreasing(increasing) sequences of simple u-functions(l-functions).[8] Their

---

[8] These functions are upper(lower) semicontinuous. Conversely, every upper (lower) semicontinuous function is a u-function(l-function).

integrals, by definition, are the limits of integrals of members in the sequence. It is proved that these limits depend only on the limit function and that in the case when some function is a u-function or an l-function, the results of both definitions agree. Before proceeding further one must show that monotonic sequences of u-functions and l-functions converging to a u-function or l-function can be integrated term by term.

We observe, of course, that the process of formation of monotonic sequences of functions previously defined can be used to extend the class of functions to which integrals can be ascribed (for example, classes ul, lu, ulu, lul, etc., are obtained). This idea is presented by Young in the form of principle (3).

(3) The principle of monotonic sequences. *A function is said to have an integral if it can be expressed as the limit of a monotonic sequence of functions whose integrals have already been defined, provided that the limit of integrals of the functions in such sequences is the same; this limit is called the integral of the given function.*

The fact that the limits of the integrals are always identical follows from Theorem 6.3 due to Young [4], [5].

Theorem 6.3

For any bounded function f obtained by some monotonic process as described above, a ul-function Ψ and lu-function ψ can be found such that $\psi(x) \leqslant f(x) < \Psi(x)$ and $\int \psi(x)\,dx = \int f(x)\,dx = \int \Psi(x)\,dx$.

In this theorem the functions ψ and Ψ depend on the monotonic sequence of functions defining f and $\int f\,dx$. However, let ψ_1, Ψ_1 be functions defined by some other monotonic sequence of functions and let $\int_1 f\,dx$ be the corresponding integral. Then $\psi_1 \leqslant \Psi$, $\Psi_1 \geqslant \psi$; integrating these inequalities we obtain $\int f\,dx = \int_1 f\,dx$; hence, the integral depends only on the limit function.

*We note that an at-most-countable collection (of step functions, systems of intervals, etc.) is used in the arguments connected with the proof of the assertions in Sections 6.1 and 6.2 describing Young's method, as well as in the proof of Theorem 6.3 for the case when f is a *ul-* or *lu-*function.

A rigorous proof of Theorem 6.3, however, as well as a rigorous formulaltion of the principle requires the use of transfinite numbers.*

Theorem 6.3 motivated Young to formulate the following definition of the integral.

Definition 6.3

Consider the integrals of all the l-functions nowhere less than f, and take the greatest lower bound of these numbers; consider similarly the integrals of all the u-functions nowhere greater than f and take the least upper bound of these numbers; if these bounds agree, then the function f is said to be integrable and the common value of these bounds is its integral.

It follows from Theorem 6.3 that all functions obtained by means of Young's principle are integrable in the sense of Definition 6.3.

Young's methods are quite similar to the geometric method of construction of Lebesgue integrals and this point is mentioned in his work. Indeed, the ordinate sets of *u*-functions (provided the points belonging to the graphs of the functions are included in these sets) and *l*-functions (without including the points of the graphs) are closed and open, respectively. Definition 6.3 corresponds to the definition of the outer measure of a set as the *infimum* of the measures of open sets containing it and its inner measure as the *supremum* of the measures of closed sets contained in it. Thus in the final result, Young's Definition 6.3 is also Lebesgue's definition,[9] formulated in terms of semicontinuous functions. However, externally, Young's definition does not utilize the notion of measure and apparently was more acceptable to mathematicians of "the older school."

If we develop Young's idea that, in Riemann's definition, integration of the given function is replaced by integration of semicontinuous functions, it is natural to arrive at the conclusion that in Lebesgue's integration procedure, the integration of the given function should be replaced by integration of *ul-* or *lu-*functions; from this point of view, Lebesgue's method is exactly one step away from Riemann's method.

---

[9] Young does not prove the equivalence of these definitions, although undoubtedly he was aware of it.

It is remarkable that additional steps are useless: by introducing into Definition 6.3 functions of the type *lul, ulu*, no additional extension of the integral is obtained; this is a corollary of Theorem 6.3.

6.7   PIERPONT'S DEFINITION

Preliminary Definition (Pierpont [2])

A separated partition D of the set A is a partition $A = \sum_i e_i$ such that the measurable covers of the sets e_i are pairwise intersecting on sets of Lebesgue measure zero and $d(e_i) \leqslant \delta$ $(\delta > 0)$.

Replacing, in his integration scheme (Chapter 2), finite partitions by countable ones, Pierpont[10] arrives at Definition 6.4 (m^*E is the outer Lebesgue measure of the set E):

Definition 6.4

Let f be a bounded function on a bounded set A. Form the sums

$$\bar{S}_D = \sum_i M_i \, m^* e_i, \qquad \underline{S}_D = \sum m_i \, m^* e_i \qquad (6.3)$$

where D is a separated partition of the set A, $M_i = \sup_{e_i} f(x)$, $m_i = \inf_{e_i} f(x)$. Then the numbers $\inf_D \bar{S}_D$, $\sup_D \underline{S}_D$ are called, respectively, the upper and the lower integrals of function f over the set A. If these integrals agree, the function f is called integrable.

A few remarks are in order to clarify the nature of this definition.

If A is measurable, then Definition 6.4 agrees with Young's first definition (Chapter 5 and Section 6.1). Indeed, if the measurable covers of the sets P and Q intersect on a set of measure zero then $m^*(P + Q) = m^*P + m^*Q$; by a limiting process this equality can be extended to a countable number of terms (the semiadditivity of the outer measure must be taken into account here). Therefore, $mA = \sum m^* e_i = m^* e_j + \sum_{i \neq j} m^* e_i$, which proves that e_i is measurable. Hence, if we ignore the inessential fact that the elements of the partitions intersect

---

[10] Volume II of Pierpont's lectures [2] was published in 1912. In the preface, Pierpont writes that this definition "occurred to him many years ago."

(on a set of measure zero), we obtain that $\bar{S}_D$ and $\underline{S}_D$ coincide with Young's sums.

*The fact that the intersection of the elements is not essential follows from Assertion 6 which explains the relation between Pierpont's and Young's procedures in the general case:

Assertion 6

For every function bounded on A there exists a measurable cover G of the set A such that the upper Pierpont integral of the function f on the set A is the infimum of the upper Young sums of the form $\sum M_i m\tilde{e}_i$ over the set G ($\{\tilde{e}_i\}$ are partitions of G) and moreover, $M_i = \sup_{A \cdot \tilde{e}_i} f(x)$.

Proof

Consider a measurable cover G^* of the set A. Let $\{\tilde{e}_i\}$, $\tilde{e}_i \cdot \tilde{e}_j = 0$ ($i \neq j$) be a partition of G^* into measurable sets, $M_i = \sup_{\tilde{e}_i \cdot A} f(x)$, $m_i = \inf_{\tilde{e}_i \cdot A} f(x)$. If we put $e_i = \tilde{e}_i \cdot A$, $D = \{e_i\}$, and take into account that $m\tilde{e}_i = m^*(\tilde{e}_i \cdot A)$, we obtain

$$\sum M_i m\tilde{e}_i \geqslant \bar{S}_D. \tag{6.4}$$

Conversely, let $\bar{S}_D$ be a sum of the form (6.3), d_i be a measurable cover of e_i, and $G^{**} \overset{\text{def}}{=} \sum d_i$ ($G^{**} \supset A$). Put $d^* = \sum_{i \neq k} d_i \cdot d_k$ and $\tilde{e}_i = d_i - d^*$; we have $md^* = 0$, $m\tilde{e}_i = m^* e_i$, $e_i \cdot e_j = 0$, $i \neq j$, $G^{**} = \sum \tilde{e}_i + d^*$. Put $\tilde{M}_i = \sup_{A \cdot \tilde{e}_i} f(x)$; since $A \cdot \tilde{e}_i \subset e_i$, $\tilde{M}_i \leqslant M_i$, and the upper Young sum $\sum \tilde{M}_i m\tilde{e}_i$ for the partition $G^{**} = \sum \tilde{e}_i + d^*$ satisfies the inequality

$$\sum \tilde{M}_i m\tilde{e}_i \leqslant \bar{S}_D. \tag{6.5}$$

Let $\{D_n\}$ be such that $\lim_{n \to \infty} S_{D_n} = \inf S_D$ and $\{G_n^{**}\}$ be the corresponding covers. It follows from inequalities (6.4) and (6.5) that $\prod_n G_n^{**}$ is the required cover G.*

If f is integrable it makes sense to proceed one step further and construct a measurable continuation φ of the function f to the region G so that Young's integral of φ over G will be equal to Pierpont's integral of f over A. This was carried out by Hildebrandt [1]. If we also utilize, following Hildebrandt, the notion of *relatively measurable subsets* of A, which are, by definition, intersections of A with L-measurable sets (this idea was used by Pierpont [1] as early as 1905! Cf. Chapter 2), we are then confronted with the following situation: relatively measurable subsets of A form a σ-ring on which the outer Lebesgue measure is additive. A bounded function f on A is integrable in the Pierpont sense if and only if the sets $E_x(f(x) > \alpha)$ are relatively measurable for any real α.

In summary, we may conclude that Pierpont's definition is a formal extension of Lebesgue's definition for the case when the region of integration is not assumed to be measurable, and from this point of view Pierpont's integral is of lesser interest.

Later, Pierpont [3] wrote that his aim was not to generalize Lebesgue's integral. In this connection, however, he drew attention to the fact that in certain situations it is more convenient to deal with an integral of a nonmeasurable function, for example, in those cases when we have to deal with cross sections of a planar set (Fubini's theorem).

6.8   LEBESGUE'S INTEGRAL AS LIMIT OF RIEMANN SUMS

Assertion 7

Let f be a summable function on $[a, b]$. Then there exists a net consisting of the sequence of partitions $\{\sigma_n\}$, $\sigma_n = \{x_i^n\}_0^{i_n}$ with diameter tending to zero, and of the points $\{\xi_i^n\}_1^{i_n}$, such that the Riemann sums $\sum_{i=1}^{i_n} f(\xi_i^n)(x_i^n - x_{i-1}^n)$ converge to the Lebesgue integral $\int_a^b f(x)\, dx$ as $n \to \infty$.

This assertion was proved by Lebesgue [15] in an even stronger form as follows: there exists a net as in Assertion 7 that works simultaneously for an arbitrary finite or countable number of summable functions. Lebesgue observes that Assertion 7 is a convenient tool for proving certain classical theorems (such as the Cauchy–Schwarz inequalities).

The proof of this assertion is immediately obtained if one first observes that given $\varepsilon > 0$, f coincides with an R-integrable function f_ε everywhere except possibly on a set of measure $\leqslant \varepsilon$, and one then takes into account the absolute continuity of Lebesgue's integral. There is no need for additional arguments in the case of a finite number of functions;

the case of a countable number of functions is treated by using the diagonal process.

However, Assertion 7 cannot be considered the foundation of a new definition of Lebesgue's integral because the prior knowledge of the Lebesgue integral is required for the construction of the "net" (see Lusin [5] in this connection). Nevertheless, Assertion 7 served as a stimulus for new searches for "regular" processes leading to the Lebesgue integral through Riemann or Darboux sums (cf. Section 6.2). Hahn [2] gives a detailed proof of the existence of a "net" for a countable sequence of functions with an *a priori* given sequence of partitions σ_n.

(It is interesting to note that Hahn proves *inter alia* the C-property of measurable functions using Vitali's theorem, which states that measurable functions agree almost everywhere with Baire functions of the second class, and does not notice that this property is proved in the very same paper by Vitali.)

In 1919, Denjoy [6] defined a regular process leading to Lebesgue's integral which outwardly is very similar to the Darboux process. He defines the *maximum* $M(f, \Delta, \alpha)$ *of a function f* on Δ with accuracy up to measure α $(0 < \alpha < 1)$, as the supremum of the numbers y such that $m E_x(f(x) > \alpha m(\Delta)$. The *minimum* $m(f, \Delta, \alpha)$ of a *function f with accuracy up to measure* α is defined as the infimum of the numbers y such that $m E_x(f(x) < y) > \alpha m(\Delta)$. Evidently, $M(f, \Delta, \alpha) \geqslant m(f, \Delta, \alpha)$ for $0 < \alpha < \frac{1}{2}$. Denjoy's definition is as follows.

Definition 6.5

Let $\sigma = \{x_i\}$ be a subdivision of the segment $[a, b]$ and $\alpha < \frac{1}{2}$. Form the sums

$$\bar{S}_\alpha(\sigma) = \sum_{i=1}^n M(f, \Delta x_i, \alpha)\, \Delta x_i, \qquad \underline{S}_\alpha(\sigma) = \sum_{i=1}^n m(f, \Delta x_i, \alpha)\, \Delta x_i.$$

The function f is called integrable if there exists a common limit to these sums as $d(\sigma)$ tends to zero.

The limit of the sums $\bar{S}_\alpha(\sigma)$, $\underline{S}_\alpha(\sigma)$ does not depend on $\alpha > 0$, i.e., it exists for *any* positive α $(\alpha < \frac{1}{2})$ if it exists for a *certain* α of this kind and its value is independent of α.

Denjoy conjectured that Definition 6.5 and Lebesgue's definition were equivalent. The proof was published in 1931 (Denjoy [7]); the summability of functions integrable in the sense of Definition 6.5 was previously proved by Kempisty [1], [2]. A number of interesting remarks concerning Definition 6.5 can be found in Lusin [5]. In the same note of 1919, Denjoy [6] also suggested two definitions of the integral more general than the Lebesgue definition. A partial analysis of these definitions was carried out by Boks [1].

7 STIELTJES' INTEGRAL

It might seem that a discussion of Stieltjes' integral would be outside the limits of the program which we set in the introduction: we agreed to consider only those methods of integration directly connected with problems of area and primitive functions. But if it is our desire to single out the essentials of Lebesgue integrations, then some details must be sacrificed; only by means of a generalization can we arrive at an understanding of what is basic and what is secondary in Lebesgue's method of integration. Stieltjes' integral is such a generalization.

7.1 HISTORICAL SURVEY

The integration methods of the 19th century are basically connected with the theory of trigonometric series. Stieltjes' integral, however, arose in a completely new and unconventional area of *continued fractions*. Remaining within the limits of this theory, this notion was a detail, hardly noticeable in the mainstream, as it was a particular generalization of the Riemann integral. This was the situation for about 15 years, until it was discovered by F. Riesz in 1909, that Stieltjes' integral was the general form of continuous linear functionals on the linear space of continuous functions. In 1910, Lebesgue published a note containing a formula expressing Stieltjes' integral of a continuous function f in terms of a Lebesgue integral of some integrable function φ of another argument. The passage from one form to the other is rather involved. Lebesgue then proposed to define Stieltjes' integral for discontinuous functions on the basis of his representation of Stieltjes' integrals; at the same time he expressed doubt as to whether there existed another sufficiently simple definition of Stieltjes' integrals of measurable functions. A few years later, Young repudiated Lebesgue's opinion by showing (in 1914) that the method of monotonic sequences applied to Stieltjes' integrals leads very simply to the same generalization.

Another approach to Lebesgue–Stieltjes integrals became possible after the notion of a *set function* was discovered. This notion became fully acceptable as a result of Lebesgue's paper [12] written in 1910; in this paper Lebesgue extends the differentiation and integration theory developed in his Leçons I for functions of a single variable to functions of several variables.

As a result of this transition to a higher dimensional space, the integral finally became envisioned as a set function; this point of view was particularly fruitful for the theory of differentiation and made it possible to single out, among various definitions of differentiation, the one in terms of which the theory acquires a unified form, independently of the number of dimensions.

In Lebesgue's investigations, set functions are still closely connected with their prototype, the integral; they are defined on the σ-ring of L-measurable sets and, provided they are "absolutely continuous," are integrals of summable functions with respect to the Lebesgue measure.

Shortly afterwards, J. Radon substantially generalized Lebesgue's arguments by considering an *a priori* given σ-ring of sets in the n-dimensional Euclidean space with an additive measure, and defined integration with respect to such a measure; he also indicated how such measures can be generated by functions of real variables. He thus constructed the theory of Lebesgue–Stieltjes integrals in a multidimensional space.

After that, all that remained was the last step in constructing an abstract theory of integrals by eliminating the requirement of specific spaces. This was done in 1915 by M. Fréchet who considered the case of a σ-ring of sets defined on an abstract space.

Up to that point, the outer measure remained an intermediate link in the structure of a measure. In 1914, C. Carathéodory considered an *a priori* given outer measure on subsets of the Euclidean space, as a nonnegative set function, satisfying the axioms of monotonicity and semiadditivity. He has shown that sets which are measurable with respect to an outer measure form a σ-ring. If in addition the metric axiom is required, then this σ-ring is not empty and contains the Borel sets.

The basic notions of modern measure theory are contained in the works of Radon, Fréchet, and Carathéodory—these are the notions of regularity and completeness connected with the extension of a measure from a ring to a σ-ring and of completion of a measure on a σ-ring.

7.2   STIELTJES' DEFINITION

*A *continued fraction* is defined by the expression

$$
a_0 + \cfrac{b_1}{a_1 + \cfrac{b_2}{a_2 + \cfrac{\ddots}{ + \cfrac{b_n}{a_n + \cfrac{b_{n+1}}{a_{n+1} + \ddots}}}}}
\tag{7.1}
$$

Such a fraction is called *convergent* if there exists a limit of *convergents* of nth order

$$
\frac{P_n}{Q_n} \overset{\text{def}}{=} a_0 + \cfrac{b_1}{a_1 + \cfrac{\ddots}{ + \cfrac{b_n}{a_n}}}
\tag{7.1'}
$$

as $n \to \infty$; this limit is taken as the value of the continued fraction.

Stieltjes [1] investigated, in his paper published in 1895, continuous fractions of the form

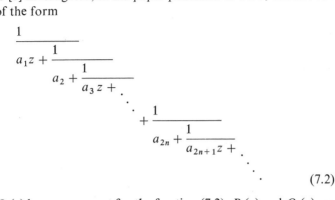

$$(7.2)$$

Let $P_n(z)/Q_n(z)$ be a convergent for the fraction (7.2); $P_n(z)$ and $Q_n(z)$ are polynomials in the complex variable z, whose coefficients depend on $a_1, \ldots, a_n$, and moreover, $Q_n(z)$ vanishes only on the negative part of the real axis. Two cases should be considered:

(1) $\sum a_n < \infty$. In this case, the convergents of even and odd orders approach different functions $q(z)/p(z)$, $q_1(z)/p_1(z)$ that have (in the finite complex plane) only simple poles as their singularities. These poles are located on the negative real semiaxis and are accumulated at $z = \infty$.

(2) $\sum a_n = \infty$. In this case, the convergents approach a unique limit; this limit is a function holomorphic outside of the negative real semiaxis; the semiaxis is generally a line of singularities in this case.

The convergent $P_n(z)/Q_n(z)$ can be expanded into the Laurent series

$$\frac{P_n(z)}{Q_n(z)} = \frac{c_0}{z} - \frac{c_1}{z^2} + \cdots + (-1)^{n-1} \frac{c_{n-1}}{z^n}$$
$$+ (-1)^n \frac{\alpha_n^{(n)}}{z^{n+1}} + (-1)^{n+1} \frac{\alpha_{n+1}^{(n)}}{z^{n+2}} + \cdots, \qquad (7.3)$$

and also into the sum of partial fractions

$$\frac{P_{2n}(z)}{Q_{2n}(z)} = \frac{M_1}{z + x_1} + \cdots + \frac{M_n}{z + x_n}, \qquad 0 < x_i < x_{i+1}, \qquad (7.4)$$

$$\frac{P_{2n+1}(z)}{Q_{2n+1}(z)} = \frac{N_0}{z} + \cdots + \frac{N_n}{z + x_n'}, \qquad 0 < x_i' < x_{i+1}',^1 \qquad (7.4')$$

^{1} The numbers x_i, x_i', M_i, and N_i depend on n.

the numbers c_i, $i = 1, \ldots, n$, are all positive and remain unchanged as n passes over to $n + 1$; the numbers M_i, N_i are also positive and, moreover, it is easy to verify that the formulas

$$c_k = \sum_{i=1}^{n} M_i x_i^k, \qquad k = 0, 1, \ldots, \tag{7.5}$$

$$c_k = \sum_{i=0}^{n} N_i x_i'^k, \qquad k = 0, 1, \ldots; \quad x_0 = 0 \tag{7.5'}$$

are valid. Expansions (7.4) and (7.4') and Eqs. (7.5) and (7.5') in the case when $\sum a_n < \infty$ retain their form in the limit as $n \to \infty$:

$$\frac{q(z)}{p(z)} = \sum_{i=1}^{\infty} \frac{M_i}{z + x_i}, \qquad \frac{q_1(z)}{p_1(z)} = \frac{N_0}{z} + \sum_{i=1}^{\infty} \frac{N_i}{z + x_i},$$

$$c_k = \sum_{i=1}^{\infty} M_i x_i^k = \sum_{i=0}^{\infty} N_i x_i'^k. \tag{7.6}$$

The numbers c_k are called the *moments* of order k of the masses M_i located at the points x_i.

In the case when $\sum a_n = \infty$, the poles of the convergents may have finite limit points and may even accumulate at each point of the negative real semiaxis; in this case expansions of the form (7.6) are generally impossible.*

Consider now, following Stieltjes, a distribution of masses over the positive real semiaxis and assume that the mass on the interval $[0, x)$ is equal to $\varphi(x)$; then the moment of the first order relative to the origin of the mass distribution on $[a, b)$ is defined as the limit of the sum

$$\sum_i \xi_i [\varphi(x_i) - \varphi(x_{i-1})],$$

as the diameter of the partition tends to zero [here $\{x_i\}$ is a partition of $[a, b)$ and $x_{i-1} \leqslant \xi_i < x_i$].

Stieltjes remarks that this limit exists also in the case when the ξ_i are replaced by $f(\xi_i)$, where f is a continuous function. He denotes this limit by

$$\int_a^b f(x) \, d\varphi(x).$$

*Consider a monotonic piecewise constant function φ with discontinuities at the points x_i at which the jump of φ is equal to M_i, $i = 1, 2, \ldots$; then the corresponding Eqs. (7.6) can be rewritten in the form

$$\frac{q(z)}{p(z)} = \int_0^\infty \frac{d\varphi(x)}{z + x}, \qquad c_k = \int_0^\infty x^k \, d\varphi(x). \tag{7.7}$$

One of the main aims of Stieltjes' investigations was to show that Eqs. (7.7) hold also in the case when $\sum a_n = +\infty$, where φ is a monotonic function (in general, strictly increasing at certain intervals). This function φ is the limit of (almost everywhere) piecewise constant functions φ_n, contructed according to Eqs. (7.4) and (7.5) for the convergents in a manner analogous to the one in which the function φ was constructed for (7.7).*

Pollard [1] in 1920 was probably the first who analyzed these definitions in detail.

7.3   RIEMANN–STIELTJES INTEGRAL—SPECIAL FEATURES

The function φ will be called henceforth the *generating function*. Usually functions of bounded variations are taken for generating functions φ; this case is then reduced to the case of a monotonic function by decomposing φ into a difference of two monotonic functions. The existence of the Stieltjes integral of a continuous function f with respect to a monotonic function φ will not be proved here; the proof, as was noted by Stieltjes, is completely analogous to the existence proof for the Riemann integral which is the special case of the Stieltjes integral with $\varphi(x) = x$. Nevertheless, there are several substantial differences between the Riemann–Stieltjes and Riemann integrals. These differences are due to the fact that in the general case the generating function is discontinuous. (There may be points for which the mass is positive, as in the case of the original Stieltjes distribution functions.)

*Several characteristic features of the Riemann–Stieltjes integrals are now presented. The reader can easily verify the statements for which a proof is not given. The function f is not assumed to be continuous.

(a)   We know that the Riemann integral can be defined in two ways: by the Riemann process as the limit of Riemann's sums and by the Darboux process as the common value of the lower and upper integrals. Can a definition of a Darboux–Stieltjes integral be given and would it be equivalent to Stieltjes' definition? The answer is negative for the reason that in the case of a discontinuous generating function, the infimum of the upper sums is not the limit of these sums; this limit as the diameter of the subdivision tends to zero generally does not exist. This characteristic as well as many other special features of Stieltjes integrals is illustrated by the following simple example:

$$f(x) = \begin{cases} 0, & 0 \leqslant x \leqslant \frac{1}{2}, \\ 1, & \frac{1}{2} < x \leqslant 1, \end{cases}$$

$$\varphi(x) = \begin{cases} 0, & 0 \leqslant x < \frac{1}{2}, \\ 1, & \frac{1}{2} \leqslant x \leqslant 1. \end{cases}$$

Here the *infimum* of the upper sums is equal to zero, since all the upper (and lower) sums are equal to zero if the point $x = \frac{1}{2}$ appears in the subdivision; however, if the discontinuity point $x = \frac{1}{2}$ is not included in the subdivision of $[0, 1]$ then the limit of the upper sums is equal to one. These "unpleasant features" are due to the fact that at the discontinuity point of φ the oscillation of f is positive. However, the property of the upper sums to vary monotonically with the addition of new points in the subdivision is preserved here. Therefore, if the limit of the upper sums exists it is equal to their greatest lower bound. In any case the following obvious proposition is valid: *in order that the Riemann–Stieltjes integral exist, it is necessary and sufficient that the limits of the extreme sums exist and that these limits be equal to one another.*

The extreme Darboux–Stieltjes integrals may coincide without implying the existence of the Riemann–Stieltjes integral.

(b)   The following theorem is proved in exactly the same manner as in the case of integrability in the Riemann sense. *In order that f be integrable with respect to a monotonic function φ on $[0, 1]$, it is necessary and sufficient that for every $\varepsilon > 0$ it be possible to include the set of discontinuities of the function f into a system of intervals $\{\delta_i\}_1^\infty$ such that $\sum_i \omega(\varphi, \delta_i) < \varepsilon$.*

This theorem was formulated by Young [6].

The following remark is in order: *Stieltjes' integral does not depend on the values assumed by the monotonic generating function at its points of discontinuity.*

(c)   We now define the Darboux–Stieltjes integral as the common value of the upper and lower Darboux–Stieltjes integrals. We have given an example which shows that such a definition is more general. This example also indicates that the Darboux–Stieltjes integral is the limit of the Darboux–Stieltjes sums if the points of discontinuity of the generating function are included in the subdivision. This fact can be interpreted as follows. Note that the Riemann–Stieltjes integrals exist on $[0, \frac{1}{2}]$ and $[\frac{1}{2}, 1]$, and nothing prevents us from *defining* the integral on $[0, 1]$ as the sum of the integrals on $[0, \frac{1}{2}]$ and $[\frac{1}{2}, 1]$. In general, if there exist Riemann–Stieltjes integrals on the segments $[a_1, a_2]$, $[a_2, a_3]$, $\ldots$, $[a_{n-1}, a_n]$, then the integral on the segment $[a_1, a_n]$ will be defined as the sum of the integrals on the segments $[a_1, a_2]$, $[a_2, a_3]$, $\ldots$, $[a_{n-1}, a_n]$. In this manner the unpleasant fact that the integral may exist on each one of the intervals without existing on their sum is avoided. However, such a definition is applicable to the case of a finite number of points $a_1, \ldots, a_n$, while the previous definition which includes the discontinuity points into the partitions is applicable to a more general case. Indeed, let $a_1, \ldots, a_n, \ldots$ be the discontinuity points of the function φ and let $n(\sigma)$ be the first index such that the point $a_{n(\sigma)}$ is not included in the partition σ. Then the following theorem is valid (Pollard [1], see also Lebesgue [4], [5]):

Theorem 7.1

The upper sums

$$S(\sigma) = \sum M_i(\varphi(x_i) - \varphi(x_{i-1})), \qquad M_i = \sup_{[x_{i-1}, x_i]} f(x),$$

approach a limit as $d(\sigma) \to 0$ *and* $n(\sigma) \to \infty$. *This limit is equal to* $\inf_\sigma S(\sigma)$, *i.e., is equal to* $\int_a^b f \, d\varphi$.

Clearly, an analogous theorem is valid for the lower sums. Hence, the Darboux–Stieltjes integral can be obtained as the limit of the Riemann–Stieltjes sums with the additional requirement that certain points should occur in the partitions. This was pointed out by Pollard who proposed to consider the Riemann–Stieltjes sum as a *multivalued function of the partition σ*.

Definition 7.1

The number L is called the limit of the partition function L(σ) if for every ε > 0, a subdivision σ(ε) can be found such that

$$|L(\sigma) - L| < \varepsilon,$$

as long as σ ⊃ σ(ε).

We note that the limit L is unique whenever it exists.

Definition 7.2   (Pollard [1])

Given φ, the function f is called an integrable function if the Riemann–Stieltjes sums L(σ) converge to L in the sense of Definition 7.1. The number L is then called the Riemann–Stieltjes integral of f with respect to φ.

Theorem 7.2   (Pollard)

The integral in the sense of Definition 7.2 is equivalent to the Darboux–Stieltjes integral.

Indeed, this theorem follows from Theorem 7.1 and the uniqueness of the limit of $L(\sigma)$.*

7.4   LINEAR FUNCTIONALS—YOUNG'S DEFINITION

As we have noted above, the Stieltjes integral received serious attention after Riesz [2] pointed out (in 1909) the connection between this integral and linear functionals.

We say that a functional $A(f)$ is defined on a certain family of functions f if the number $A(f)$ corresponds to each function f. Riesz considered the family $C[0, 1]$ of all continuous functions on the segment $[0, 1]$ and a function $A(f)$ on this family satisfying two conditions:

   (a)   *Linearity*: $A(f_1 + f_2) = A(f_1) + A(f_2)$.
   (b)   *Continuity*: $A(f_n) \rightarrow A(f)$, if the sequence $\{f_n\}$ tends uniformly to f.

The following theorem is proved by Riesz.

Theorem 7.3

Every functional $A(f)$, linear and continuous on the space $C[0, 1]$, can be represented as

$$A(f) = \int_0^1 f \, d\varphi,$$

where φ is a function of bounded variation on $[0, 1]$ [depending on $A(f)$].

In 1910, Lebesgue [11] derived a formula reducing the Stieltjes integral to the Lebesgue integral. This formula is of the form

$$\int_a^b f(x) \, d\varphi(x) = \int_\alpha^\beta f(x(v)) \frac{d\varphi(v)}{dv} \, dv, \tag{7.8}$$

where $v(x)$ is the total variation of the function φ on $[a, x]$ and $x(v)$ is the inverse function of $v(x)$. [This function is appropriately defined in case v possesses intervals of constancy.]

Lebesgue notes that Eq. (7.8) can be used to define the Stieltjes integral of a discontinuous function f as the Lebesgue integral in the right-hand side of (7.8); since $|d\varphi/dv| = 1$ almost everywhere, bounded measurable functions of argument v are integrable. Lebesgue believed at that time that it would be difficult to generalize Stieltjes' integration by any other means. However, three years later Young [6] showed that a very natural method for such a generalization exists, based on the method of monotonic sequences.

The process for constructing an integral using a given generating function φ, described by Young, follows the procedure presented in Chapter 6. It is only necessary to define properly the integral $\int_a^b f(x) \, d\varphi(x)$ of a step function f. Young assumes that the points of discontinuity of the function f are included among the partition points x_i, $a < x_0 < \cdots < x_n < b$.

Then

$$\int_a^b f \, dg \stackrel{\text{def}}{=} \sum_i \{f(c_i)[g(c_i + 0) - g(c_i)] + f(c_i + 0)[g(c_{i+1} - 0)$$

$$- g(c_i + 0)] + f(c_{i+1})[g(c_{i+1}) - g(c_{i+1} - 0)]\}. \tag{7.9}$$

We note that in general f is integrable neither in the Riemann–Stieltjes sense nor in the Darboux–Stieltjes sense; however, one can easily guess the origin of the sum (7.9) which agrees with the concept of the Stieltjes integral as a sum of masses: indeed, it is constructed from the masses of rectangles with base (c_i, c_{i+1}) and constant height $f(c_i + 0)$ (the middle term in the braces) and also from the masses of the ordinates $f(c_i)$ equal to $f(c_i)[g(c_i + 0) - g(c_i - 0)]$. (Each one of these masses is subdivided into two parts: $f(c_i)[g(c_i + 0) - g(c_i)]$ and $f(c_i)[g(c_i) - g(c_i - 0)]$; in Eq. (7.9), one of these masses appears for the point c_i and the other for the point c_{i+1}.)

Having integrals of simple u- and l-functions, we define the integrals of u- and l-functions (and then of lu- and ul-functions) successively by means of a monotonic limiting process. Definition 6.3 (p. 100) serves as a definition of an integral of an arbitrary (bounded) function; moreover, all the bounded functions in Young's classification turn out to be integrable. For the unbounded functions the integral is defined by means of de la Vallée-Poussin's procedure.

Young shows that for the case of a function f integrable in the Riemann–Stieltjes sense, his definition is equivalent to Stieltjes' definition and he formulates the corresponding necessary and sufficient criterion of integrability in the form presented in Section 7.3 of this chapter. He remarks that this criterion can be formulated in terms of a measure analogously to the formulation of the integrability condition in the Riemann sense, where the measure is replaced in this case by the infimum of the sums of the oscillations of the function φ over the systems of intervals containing the set.

Developing his idea, Young points out that for the case of an arbitrary set, this lower bound is an analog to the notion of the outer measure and he calls it *the upper variation of the function φ over the set*. The variation of the function φ on a set E can be defined as the integral $\int_a^b \chi_E \, d\varphi$ of the characteristic function of the set E; if the integral does not exist the upper and lower integrals (in the sense of Definition 6.3) represent the upper and lower variation of the function φ on the set E.

The notions of variation, upper variation, and lower variation of a function φ correspond to the notions of Lebesgue's outer and inner measures, obtained for $\varphi(x) = x$. However, Young confines himself to the above remarks and does not deal with the construction of the theory of φ-measurable functions. A construction of a theory of this type was carried out by Radon.

7.5 SET FUNCTIONS

As we have already noted at the beginning of this chapter, Lebesgue develops the integration and differentiation theory for the case of an n-dimensional Euclidean space in the paper "Sur l'intégration des fonctions discontinues" [12]. We shall not dwell on the analogies, but concentrate on the new concepts introduced by Lebesgue in this work; one of these is undoubtedly the notion of set function.

In an effort to single out the most general properties of integrals which do not depend on the dimensionality of the space, Lebesgue obtains that the integral[2] $\int_E f$ for a given function f is a set function $\Phi(E)$ defined on a σ-ring of L-measurable sets E, possessing the following fundamental properties:

(1)   The function is *countably additive*, i.e., if $\{E_i\}$ is a sequence of pairwise nonoverlapping L-measurable sets, then

$$\Phi\left(\sum_i E_i\right) = \sum_i \Phi(E_i).$$

(2)   The function is *absolutely continuous*, i.e., $\Phi(E) = 0$, if $mE = 0$.
(3)   The function is of a *bounded variation*, i.e., $|\Phi(E)| < M$ for any E.

The numbers

$$\overline{W}(\Phi, E) \overset{\text{def}}{=} \sup_{\mathscr{E} \subset E} \Phi(\mathscr{E}),$$

$$\underline{W}(\Phi, E) \overset{\text{def}}{=} - \inf_{\mathscr{E} \subset E} \Phi(\mathscr{E}),$$

$$W(\Phi, E) \overset{\text{def}}{=} \overline{W} + \underline{W}$$

are called, respectively, the *upper*, *lower*, and *absolute* variation of the set function $\Phi(E)$. Since $\Phi(E)$ by definition equals zero on an empty set, we have $-\underline{W}(\Phi, E) \leqslant 0 \leqslant \overline{W}(\Phi, E)$.

Properties (1)–(3) are not independent since, as Lebesgue points out, a *countably additive set function is a function of bounded variation*.

---

[2] The symbol $\int_E f$ denotes here an n-fold integral.

*In order to familiarize the reader with the notion of set functions without reference to special literature on this subject (such as Saks [1]), we present here the proof of the above statement.

Let $\Phi(E)$ be defined on T. First note that if $\{E_n\}$ is a monotonic sequence of sets then $\lim \Phi(E_n) = \Phi(\lim E_n)$. Assume now that $W(\Phi, T) = +\infty$. Note that if the function $\Phi(E)$ is of unbounded variation on $\tilde{E}$ (this means that $\sup_{E \subset \tilde E} |\Phi(E)| = +\infty$), then for every n there exists a subset $E_1 \subset \tilde{E}$ such that $|\Phi(E_1)| > n$ and the variation of $\Phi(E)$ is unbounded on E_1. Indeed, let the set $\mathscr{E} \subset \tilde{E}$ be such that $|\Phi(\mathscr{E})| > n + |\Phi(\tilde{E})|$, then $|\Phi(\tilde{E} - \mathscr{E})| > n$ [since $\Phi(\mathscr{E}) + \Phi(\tilde{E} - \mathscr{E}) = \Phi(\tilde{E})$], and the variation of $\Phi(E)$ is unbounded on one of the sets $\mathscr{E}$ or $\tilde{E} - \mathscr{E}$. This set is taken as E_1. We now define by induction a monotonically decreasing sequence of sets E_n such that $|\Phi(E_n)| > n$; but then $\lim \Phi(E_n) = \Phi(\lim E_n) = \pm\infty$, which is impossible. Hence, $|\Phi(E)| < \infty$.   Q.E.D.

Clearly all the variations of an absolute continuous set function are also absolutely continuous set functions.

We now establish the formula

$$\Phi(E) - \overline{W}(\Phi, E) - \underline{W}(\Phi, E) \tag{7.10}$$

due to Lebesgue.

Let the set $\mathscr{E} \subset E$ be such that $\Phi(\mathscr{E}) > \overline{W}(\Phi, E) - \varepsilon$. Write $\Phi(E) = \Phi(\mathscr{E}) + \Phi(E - \mathscr{E}) \geqslant \overline{W}(\Phi, E) - \varepsilon - \underline{W}(\Phi, E)$; since ε is arbitrary, we obtain $\Phi(E) \geqslant \overline{W}(\Phi, E) - \underline{W}(\Phi, E)$. The reverse inequality is obtained by defining ε so that $\Phi(\mathscr{E}) < -\underline{W}(\Phi, E) + \varepsilon$.

The decomposition (7.10) is analogous to a well-known decomposition of a function of bounded variation of a real variable into the difference of two monotonic functions.*

The integral $\int_E f\, dx$ is a classical example of a countably additive absolutely continuous set function and, moreover,

$$\overline{W}(\Phi, E) = \int_E f^+\, dx, \qquad \underline{W}(\Phi, E) = \int_E f^-\, dx.$$

Lebesgue's remarkable result is as follows: *every countably additive set function, absolutely continuous on L-measurable sets, is a Lebesgue integral.*

This theorem is preceded by a discussion of the derivatives. Every absolutely continuous countably additive function $\Phi(E)$ is the integral of its derivative (which exists almost everywhere). For the case of a segment on the real line, this derivative is the usual derivative of the absolutely continuous function $\Phi([a, x])$ of the real variable x. Lebesgue introduces a more general form of differentiation which is applicable to a set function in the n-dimensional Euclidean space; in view of the importance of this new notion, we present here the corresponding definitions (in the n-dimensional space).

Definition 7.3

A sequence $\{E_n\}$ of measurable sets is called regularly contracting to the point x if: (1) $x \in E_n$ for every n; (2) $\lim_{n \to \infty} d(E_n) = 0$; (3) there exists a positive number α such that $(mE_n/mT_n) > \alpha$, where T_n is the smallest "cube" containing the set E_n.

Definition 7.4

If for any sequence of measurable sets $\{E_n\}$ regularly contracting to the point x, there exists the limit

$$\lim_{n \to \infty} \frac{\Phi(E_n)}{mE_n} = \Phi'(x)$$

independently of the sequence, then $\Phi'(x)$ is called the derivative of the set function Φ at point x.

Here the condition of regularity is necessary; as Lebesgue points out, without this condition the multiple integral would be differentiable only at those points at which the integrand is continuous, if its values on a set of measure zero are neglected. This remains valid even if we assume that E_n are domains. Finally, if the choice of E_n becomes even more restricted and they are assumed to be rectangles with sides parallel to the axis, the condition of regularity remains nevertheless necessary. This case was not fully evident to Lebesgue at that time ([12], p. 303, footnote; details can be found in Saks' book [1]).

Lebesgue utilizes the method of *chains of intervals* in the proof of Theorem 7.1. This method cannot be extended in a natural manner to the case of n-dimensional space; it is replaced by Vitali's covering theorem (see e.g., Natanson [1]). Lebesgue formulates this theorem in the general form: *Let* $\{E_\alpha\}$ *be a system of measurable sets whose sum contains a measurable set E, and let a sequence of sets belonging to the system* $\{E_\alpha\}$, *regularly contracting to this point, correspond to each point x of the set E. Then there exists an at most countable sequence* $\{E_n\}$ *of pairwise nonoverlapping sets of the system* $\{E_\alpha\}$ *whose sum contains almost all the set E.*

Along with set functions one can also consider domain functions $\Phi(D)$ defined on quadrable domains D.

A quadrable domain is a domain whose boundary has Lebesgue measure zero. i.e., the domain is measurable in the Jordan sense. Lebesgue requires that an *additive domain function*[3] be defined by the relation $\Phi(D) = \sum \Phi(D_i)$ if the domain D is the sum of a closed domains D_i with pairwise disjoint interiors. Such a function is of bounded variation if the sums $\sum_i |\Phi(D_i)|$ are uniformly bounded, where $\{D_i\}$ is a sequence of closed domains contained in D, with pairwise disjoint interiors. Absolute continuity of $\Phi(D)$ means that for every $\varepsilon > 0$, a $\delta > 0$ can be found such that $\left| \sum \Phi(D_i) \right| < \varepsilon$, provided $\sum mD_i < \delta$.

Lebesgue shows how one passes from an additive absolutely continuous domain function to a set function: to obtain $\Phi(E)$ one must consider the coverings of the set E by means of disjoint domains $\{D_i\}$ such that $\sum_i mD_i - mE < \varepsilon$. Then $\Phi(E) \overset{\text{def}}{=} \lim_{\varepsilon \to 0} \sum_i \Phi(D_i)$.

Lebesgue devotes a great deal of attention to the investigations of additive functions of rectangles $(x_1 < x < x_2, y_1 < y < y_2)$ that are of bounded variation, but not necessarily absolutely continuous; such a function can be obtained by means of functions $\Psi(x, y)$, putting

$$\Phi([x_1, y_1, x_2, y_2]) = \Psi(x_2, y_2) - \Psi(x_1, y_2)$$
$$- \Psi(x_2, y_1) + \Psi(x_1, y_1). \tag{7.11}$$

The following observations due to Lebesgue are required for the text to follow:

(1)   when we define an additive set function, we define at the same time a function of rectangle;

---

[3] Either countable or finite additivity can be assumed here.

(2)   this function of rectangle is generally not additive, since the set function (not absolutely continuous in general) may not be equal to zero on a side of the rectangle.

We note in passing that Lebesgue does not indicate how the reverse process, namely, the passage from a function of rectangle of bounded variation to a set function, can be accomplished.

Thus, the characteristic features of the paper "Sur l'intégration des fonctions discontinues" (Lebesgue [12]) are as follows:

(a)   additive set functions are defined on a σ-ring of L-measurable sets;

(b)   these functions are absolutely continuous and are therefore Lebesgue integrals;

(c)   the connection between the domain (rectangle) functions and the set functions is pointed out under the condition of absolute continuity of these functions; it is not shown however how one obtains in general a set function from an additive function of rectangle (of bounded variation).

7.6   RADON'S INTEGRAL

Radon's investigations (Radon [1], 1913) are a direct continuation and generalization of Lebesgue's results discussed above. His main contributions are as follows:

(a')   Additive set functions[4] are defined on an arbitrary *a priori* given σ-ring T of sets in n-dimensional Euclidean space.

(b')   If $b(E)$ is a nonnegative additive function defined on T (a *basis* in Radon's terminology), then relative to this function (called a measure), a class of absolutely continuous functions is singled out. In this case the following theorem holds: *An additive function absolutely continuous relative to $b(E)$ is an integral with respect to the measure $b(E)$.*

(c')   Functions of a *half-open* rectangle are considered which possess the property of countable additivity: As a result of this, the interrelation between functions of rectangles and set functions on one hand and real-variable functions on the other becomes simple.

---

[4] Henceforth, countably additive functions will be called additive.

Furthermore, outer measures, obtained by means of functions of rectangles, and classes of sets measurable relative to the outer measure are investigated.

We shall now examine the specific features of Radon's papers in more detail. For the sake of brevity we restrict our discussion to a half-closed interval $[-M, M)$ in R_1.

Assume (as is done by Radon) that the domain of definition (*Definizionbereich*) of the set function $\Phi(E)$ is a σ-ring of sets T, containing half-intervals of the form $[a, b)$; hence, T contains B-measurable sets. As above, three variations of set functions are defined, and, moreover, the expansion (7.10) is valid. Therefore, it is sufficient to consider only nonnegative set functions $\Phi(E)$. Define

$$F(x) \stackrel{\text{def}}{=} \Phi([=M, x)), \qquad F(-M) \stackrel{\text{def}}{=} 0;$$

F is a monotonically increasing function. The function of a half-closed interval $\Phi([a, b))$ is connected with F by the relation analogous to relation (7.11):

$$\Phi([a, b)) = F(b) - F(a). \tag{7.12}$$

Let $\{x_n\}$ be a monotonic sequence which approaches x from the left; then in view of the countable additivity of the function Φ, we obtain $\lim_{x_n \to x} \Phi([-M, x_n)) = \Phi([-M, x))$ or

$$\lim_{n \to \infty} F(x_n) = F(x). \tag{7.13}$$

The function F is continuous from the left. The main result of Radon in connection with set functions is the proof of the invertability of Eq. (7.12); namely, the validity of Theorem 7.4.

Theorem 7.4

Every monotonically increasing real function F satisfying Eq. (7.13) defines an additive set function $\Phi(E)$ on some σ-ring of sets containing half-closed intervals. Moreover, Eq. (7.12) holds.

We sketch the proof of this theorem. Using the monotonic function F, we define a function of half-interval $\Phi([a, b))$ using Eq. (7.12). This function is finitely additive on half-closed intervals (i.e., if $[a, b) = [a, b_1) + [b_1, b_2) + \cdots + [b_{n-1}, b)$, $b_i < b_{i+1}$, then $\Phi([a, b)) = \Phi([a, b_1)) + \Phi([b_1, b_2)) + \cdots + \Phi([b_{n-1}, b))$; moreover, the condition of continuity from the left given by (7.13) assures the countable additivity of Φ:

$$\Phi([a, b)) = \sum_{n=1}^{\infty} \Phi([b_{n-1}, b)), \qquad b_0 = a, \quad b_n \uparrow b, \qquad (7.14)$$

or, equivalently, the continuity of Φ as a function of half-closed intervals:

$$\lim_{n \to \infty} \Phi([a, b_n)) = \Phi([a, b)), \qquad b_n \uparrow b. \qquad (7.15)$$

Next, an outer measure is defined on the subsets $E \subset [-M, M)$ using the standard procedure

$$\Phi^*(E) = \inf \sum \Phi([a_n, b_n)), \qquad \sum_n [a_n, b_n) \supset E.$$

This outer measure is easily verified to be a nonnegative, monotonic, and subadditive set function, and in view of condition (7.14) [or (7.15)], $\Phi^*([a, b)) = \Phi([a, b))$. Thus $\Phi^*(E)$ possesses the basic properties of the Lebesgue outer measure m^*E; (the proof of these properties is carried out in the same manner as in the case of m^*E.[5] Φ-measurable sets E are determined by the relation

$$\Phi^*(E) + \Phi^*(CE) = \Phi([-M, M)).$$

As in the case of the Lebesgue measure it is proved that the Φ-measurable sets form a σ-ring T^* which includes the Borel sets; the outer measure $\Phi^*(E)$ on this σ-ring is a nonnegative countably additive set function (i.e., a measure). This is the required function $\Phi(E)$ in Theorem 7.4.

[5] To preserve these properties, property (7.13) [or (7.15)] is essential, since it guarantees the countable additivity of a finite-additive function Φ viewed as a function of a half-closed interval: the fact of the matter is that finite additivity and continuity are equivalent to countable additivity. To understand these facts more clearly, we recommend that the reader define by means of Eq. (7.12) functions of segments $\Phi([a, b])$ and of intervals $\Phi((a, b))$ and check whether or not the constructed set function $\Phi^*(E)$ is additive and continuous.

The measure $\Phi(E)$ possesses the following property: for any measurable set E and any $\varepsilon > 0$, an open set G containing E can be found such that

$$\Phi(G) - \Phi(E) < \varepsilon; \qquad (7.16)$$

the existence of such a set G follows directly from the definition of the outer measure.

Remark. The property expressed by relation (7.16) observed by Radon, is equivalent to the following: for every measurable set E there exists a B-kernel and a B-cover (see Section 4.2). It follows from here that each set of the σ-ring T^* differs from some Borel set on a set Φ-measure zero (which is a subset of some Borel set of Φ-measure zero).

The following question was posed by Radon: let a set function $\Phi(E)$ on a σ-ring T, containing half-closed intervals be given; first an outer measure $\Phi^*(E)$ and then the σ-ring T^* of measurable [relative to $\Phi^*(E)$] sets is constructed by means of its values on half-closed intervals; what is the relation of T^* to T? In the common part of T and T^*, Borel sets are included in any case; if $E \subset T \cdot T^*$, then $\Phi(E) = \Phi^*(E)$; this is about all that can be said in general. However, as Radon points out, if $\Phi(E)$ possesses the property expressed by relation (7.16) on T, then $T \subset T^*$. Radon calls the σ-ring T^* *the natural domain of definition* of $\Phi(E)$ (*natürlicher Definizionbereich*).

We note that T^* depends only on the values of Φ assumed on the half-closed intervals.

The natural domain of definition T^* for the case of a general set function $\Phi(E)$ is defined as the intersection of the σ-rings T^* for $\overline{W}(\Phi, E)$ and $\underline{W}(\Phi, E)$.

Knowing how to construct a countably additive measure from a given monotonic function F, we can, using Radon's method, generalize Stieltjes' integral in the same manner as Lebesgue generalizes Riemann's integral. We note only that by changing the value of F at the discontinuity points, one can always assume that condition (7.13) is satisfied. (As we have seen above, this change has no effect on the Riemann–Stieltjes integral.)

Therefore, there is no need to dwell on the construction of the Lebesgue–Stieltjes integral (it is also called the Radon integral) since the procedure is completely analogous to the one used in the construction of Lebesgue's integral; in this case, Lebesgue's measure is replaced by the Φ-measure (i.e., by a nonnegative additive set function on a σ-ring of sets which are called measurable sets) and the measurable functions by Φ-measurable functions, i.e., functions f for which the sets $E_x(f(x) > a)$ are Φ-measurable for any real a. The Lebesgue–Stieltjes integral is the common limit of the upper and lower sums

$$\Sigma = \sum_{-\infty}^{+\infty} y_k \Phi\left(E_x(y_{k-1} \leqslant f(x) < y_k)\right),$$

$$\sigma = \sum_{-\infty}^{+\infty} y_{k-1} \Phi\left(E_x(y_{k-1} \leqslant f(x) < y_k)\right).$$

It is easy to see that in the case of a step function f, the Lebesgue–Stieltjes integral $\int_a^b f(x)\, d\varphi(x)$ is calculated using Eq. (7.9). Since monotonic sequences of functions can be integrated term by term, we conclude that the Radon and Young integrals of ul- and lu-functions agree; complete equivalence of Radon's and Young's definitions holds when Φ is considered in the natural domain of definition (more details are given in Section 7.7) and this follows from arguments given in Section 6.6, which establish this equivalence for the case of an integral with respect to the Lebesgue measure and which remain valid for the case of the Φ-measure and Φ-measurable functions.

Let $\Phi(E)$ and $b(E)$ be additive set functions defined, respectively, on the σ-rings T and $\tilde{T}$, containing the Borel sets and, moreover, let $b(E) \geqslant 0$. The function $b(E)$ is called by Radon a *basis*[6] for $\Phi(E)$ if $b(E) = 0$ implies $\Phi(E) = 0$ for any Borel set E. This condition is equivalent to the following: for any $\varepsilon > 0$, a $\delta > 0$ can be found such that $b(E) < \delta$ implies $|\Phi(E)| < \varepsilon$. Clearly, we are dealing here with a generalization of the notion of an absolutely continuous set function to the case of a measure $b(E)$. From the remark stated on p. 124, it follows that the natural domain of definition of $\Phi(E)$ contains the natural domain of definition of $b(E)$. Moreover, the following generalization of Lebesgue's theorem was proved by Radon: *the function $\Phi(E)$ on the natural domain of definition of the function $b(E)$ is an integral $\int_E f(x)\, db(E)$ with respect to the measure $b(E)$.*

The final step in the generalization of the Lebesgue integral was carried out by Fréchet. This is discussed in Section 7.7.

[6] See also the beginning of this section.

7.7   INTEGRALS IN ABSTRACT SPACES

The new element appearing in Fréchet's paper (Fréchet [1], 1915) is the introduction of an *abstract space*, i.e., a set F of elements of an arbitrary nature. Specific properties of elements of F are not essential for the following discussion.

In the space F, constructions are carried out based on principles already familiar from the works of Lebesgue and Radon. More specifically, a σ-ring of sets T with a given countably additive set function $\Phi(E)$ is defined in F. The variations $\overline{W}(\Phi, E)$ and $\underline{W}(\Phi, E)$ of the function $\Phi(E)$ are defined in the standard way and the decomposition (7.10) is established.

Fréchet discusses in some detail the particular features of integration in abstract spaces. Let f be a nonnegative function of an element $x \in F$. The function f is called summable (integrable) in Fréchet's sense (relative to Φ) on the set $E \subset T$ if the series $\sum_i M_i W(\Phi, E_i)$ is convergent for any partition of the set E into nonoverlapping components E_i, $E_i \in T$, $M_i = \sup_{E_i} f$. *The integral in the Fréchet sense* (F-integral) of such a function is defined as the difference of the integrals relative to $\overline{W}(\Phi, E)$ and $\underline{W}(\Phi, E)$, and these two integrals are defined following Young's definitions as the common value of the upper and lower integrals.

The reader can easily verify (without referring to Lebesgue's definition!) that the integral defined in this manner possesses properties (2), (3), and (4) of an integral of a summable function presented at the end of Section 4.5, and that one may integrate term-by-term uniformly convergent sequences of functions.

A question arises concerning the equivalence of the definition given above and the definition based on Lebesgue's procedure for the case of an abstract σ-ring. In the general case, Theorem 7.5 due to Fréchet is valid.

Theorem 7.5

In order that a function f be integrable in the Fréchet sense, it is necessary and sufficient that, for every real a, the set $E_x(f(x) > a)$ differs from a set in the σ-ring T by a subset of a set of measure zero belonging to the σ-ring.

*We now prove this theorem. Assume that $\Phi(E)$ is nonnegative and that f is integrable with respect to Φ on the set E. We can assume that

$$\text{(F)} \int_E f(x)\,dx = \lim_{k\to\infty} \sum_i M_i^k \Phi(E_i^k) = \lim_{k\to\infty} \sum_i m_i^k \Phi(E_i^k),$$

where $\{E_i^k\}$ are monotonically decreasing systems of subdivisions of the set E, i.e., it follows from $E_i^{k+1} \cdot E_j^k \neq 0$ that $E_i^{k+1} \subset E_j^k$. Under this assumption, the functions

$$\Psi_k(x) \stackrel{\text{def}}{=} \sum_i M_i^k \chi_i^k(x), \qquad \psi_k(x) \stackrel{\text{def}}{=} \sum_i m_i^k \chi_i^k(x),$$

where $\chi_i^k(x)$ is the characteristic function of the set E_i^k, form monotonic sequences and

$$\psi(x) = \lim_{k\to\infty} \psi_k(x) \leqslant f(x) \leqslant \lim_{k\to\infty} \Psi_k(x) = \Psi(x).$$

Moreover, in view of the integrability of f,

$$\text{(F)} \int_E \psi(x)\,dx = \text{(F)} \int_E \Psi(x)\,dx.$$

Hence, $\psi(x) = f(x) = \Psi(x)$ on E except possibly on a set e of Φ-measure zero. Now the functions ψ_n, Ψ_n and also ψ and Ψ are measurable[7]; hence, f is measurable on $E - e$ and the set $\mathscr{E}_{x \in E}(f > a)$ differs from the measurable set $\mathscr{E}_{x \in E - e}(f > a)$ by the subset $\mathscr{E}_{x \in e}(f > a)$ of a set e of zero measure. The necessity of the condition of the theorem is proved; the sufficiency follows from the fact that the F-integral is contained between the upper and lower Lebesgue sums.*

Definition 7.4′

A σ-ring T is called complete relative to a measure Φ defined on it if T contains all the subsets of sets of measure zero.

Fréchet points out that if the sets of the form

$$E + e' - e''$$

[7] Measurability means here that the sets $\mathscr{E}_x(f(x) > a)$ belong to the σ-ring T.

are added to the σ-ring T, where $E \subset T$ and the sets e' and e'' are subsets of sets of measure zero belonging to T, and if the measure Φ is defined on these sets by the equality

$$\Phi(E + e' - e'') \overset{\text{def}}{=} \Phi(E), \tag{7.17}$$

then, as it is easy to verify, a new σ-ring is obtained which is complete relative to the measure Φ. The same σ-ring is obtained if the functions

$$\Phi^*(E) = \inf_{\mathscr{E} \supset E, \, \mathscr{E} \subset T} \Phi(\mathscr{E}), \qquad \Phi_*(E) = \sup_{\mathscr{E} \subset E, \, \mathscr{E} \subset T} \Phi(\mathscr{E})$$

are constructed and the sets E for which $\Phi^*(E) = \Phi_*(E)$ are singled out.

7.8   CARATHÉODORY'S MEASURE

Studying the development of the Stieltjes integral we were able to detect a close connection between this notion and the notion of measure. Moreover, one may say that this development was predetermined to a great extent by the evolution in the notion of measure. Every constructively defined measure is an outer measure whose values are considered on a σ-ring on which the measure is additive. The existence of such a σ-ring is a corollary of certain properties of the outer measure; this was noted by Carathéodory [1] who formulated these properties as axioms. A set function in the space R_n, which satisfies these axioms is called the *outer Carathéodory measure* (see, e.g., Saks [1], Chapter II).

Definition 7.5

A set function $\Phi^*(E)$ defined on subsets of an n-dimensional Euclidean space is called an outer measure if it satisfies the following conditions:

(1) $\Phi^*(E)$ is nonnegative and equals zero on the empty set; the case $\Phi^*(E) = +\infty$ is not excluded.

(2) $\Phi^*(E)$ is monotonic, i.e., $\Phi^*(E_1) \leqslant \Phi^*(E_2)$ if $E_1 \subset E_2$.

(3) $\Phi^*(E)$ is subadditive, i.e., if $E = \sum_i E_i$, then $\Phi^*(E) \leqslant \sum_i \Phi^*(E_i)$.

These properties of the outer measure are sufficient to single out the class of sets measurable relative to $\Phi^*(E)$.

Definition 7.6

The set A is called measurable if for any set M the equality.

$$\Phi^*(M) = \Phi^*(M \cdot A) + \Phi^*(M \cdot CA) \tag{7.18}$$

is satisfied.

This definition corresponds completely to Lebesgue's and Young's definitions; it is required that Eq. (7.18) hold for all M since Φ may possibly be infinite on some sets. If Φ is finite, then from the definition of measurability, for example, on subsets of a segment $[a, b]$, it is sufficient to require that (7.18) be satisfied for $M = [a, b]$. This is the case in Lebesgue's definition.

Carathéodory shows that the collection of measurable sets form a σ-ring on which Φ is countably additive. However, if the outer measure is defined in such a general form, one cannot prove that the class of measurable sets is nontrivial! To guarantee existence of nonempty measurable sets different from the whole space, Carathéodory stipulates the following additional requirement:

(4) $\Phi(E_1 + E_2) = \Phi(E_1) + \Phi(E_2)$ if $d(E_1, E_2) > 0$.

If this requirement is satisfied, then the open sets will be measurable; hence, all Borel sets will be measurable. An outer measure which satisfies requirement (4) is called a *metric outer measure*.

Carathéodory singles out a class of measures which he calls *regular*. These satisfy the following requirement:

(5)   For any set E, $\Phi^*(E)$ is the greatest lower bound of the numbers $\Phi^*(\mathscr{E})$, where $\mathscr{E}$ is an arbitrary measurable set containing E.

The most important outer measures are regular; in particular, such are the outer measures considered by Radon. If the measure $\Phi^*(E)$ is regular, then as it was pointed out by Carathéodory, an inner measure $\Phi_*(E)$ can be defined as the least upper bound of the measures of the measurable subsets of the set E, and it possesses all the usual properties of the Lebesgue inner measure; for the measurability of a set E, it is necessary and sufficient that

$$\Phi^*(E) = \Phi_*(E).$$

III INTEGRATION IN THE SECOND DECADE OF THE 20TH CENTURY

8 THE PROBLEM OF THE PRIMITIVE —THE DENJOY–KHINCHIN INTEGRAL

As we have seen in Chapter 4, Lebesgue's integral reproduces the primitive of an exact finite summable derivative[1]; there are, nevertheless, continuous functions whose exact derivative is not summable. Thus the problem of the determination of a primitive of an exact finite derivative is in general insoluble in the framework of the Lebesgue theory. The problem remained unsolved until 1912 when Denjoy, in two notes published in Volume 154 of *Comptes Rendus,* presented the final solution to the problem of the primitive. Before discussing his results, however, we shall consider some particular cases noted by Lebesgue as early as 1902.

---

[1] When referring to the exact derivative of a function F, we assume that F has a derivative at each point and that the derivative may be finite or infinite. We note, however, that in general the function is not determined uniquely by its exact derivative if the derivative is infinite at certain points (see Lebesgue [4], [5]).

8.1 PRELIMINARY RESULTS

In ILA, Lebesgue [2] considers the case when the set of points, at which the derivative is nonsummable,[2] is reducible and points out that in this case Dirichlet's methods enable us to determine the primitive. However, this method is clearly not applicable in the case when the set of points at which the derivative is nonsummable is of a positive measure; an example of such a derivative can easily be constructed. Let P be a perfect set on $[a, b]$, $CP = \sum \delta_i$. Consider the function F equal to zero on P and equal to

$$(x - \alpha_n)^2(x - \beta_n)^2 \sin \frac{1}{(x - \alpha_n)^2(x - \beta_n)^2}$$

on each adjacent interval $\delta_n = (\alpha_n, \beta_n)$. F is obviously differentiable in the interior of every interval δ_n. We show that $F' = 0$ at each point $x_0 \in P$. The difference ratio

$$r(F, x_0, x_0 + h) = \frac{F(x_0 + h) - F(x_0)}{h}$$

is equal to zero for $x_0 + h \in P$; if $x + h \in \delta_n$, we have $h > (x_0 + h) - \alpha_n$ for $h > 0$ and $-h > \beta_n - (x_0 + h)$ for $h < 0$. Therefore, in both cases

$$|r(F, x_0, x_0 + h)| \leqslant |x_0 + h - \alpha_n| \cdot |x_0 + h - \beta_n|;$$

the right-hand side of this inequality tends to zero as $n \to \infty$, i.e., $F'(x_0) = 0$. It is directly verified that F' is nonsummable at the end points of δ_n; hence, every point of the set is a point at which F' is not summable.

In this example F' is summable on the set P of the nonsummability points. This of course is not always true; but if this is the case and if additionally the series $\sum_i (F(\beta_i) - F(\alpha_i))$ is absolutely convergent, then,

---

[2] Let α be a real number. Suppose there exists $\beta > \alpha$ ($\beta < \alpha$) such that for each γ, $\alpha < \gamma < \beta$ ($\beta < \gamma < \alpha$), the function f is summable on $[\gamma, \beta]$ ($[\beta, \gamma]$), but is *not* summable on $[\alpha, \beta]$ ($[\beta, \alpha]$). Then α is called *a point of nonsummability* of f. (*Translator's note.*)

as it was noted by Lebesgue, the primitive is determined by the following formula

$$F(x) - F(a) = \sum_{\delta_i < x} (F(\beta_i) - F(\alpha_i)) + \int_{P \cdot [a, x]} F'(x) \, dx.^3 \qquad (8.1)$$

We shall see in what follows that by establishing Eq. (8.1), Lebesgue made the first step on the transfinite path which led to the determination of the primitive as suggested by Denjoy.

We shall now prove Eq. (8.1) using Lebesgue's method; this formula will allow us to simplify certain further arguments.

Let $P_{x_0} = P \cdot [a, x_0] + \{x_0\}$ ($\{x_0\}$ is the set consisting of a single point). Consider a continuous function G defined on $[a, x_0]$ which is equal to F on P_{x_0} and which is linear on the intervals $\delta_i = (\alpha_i, \beta_i)$ adjacent to P_{x_0}. The function G is differentiable not only at each point $x \in \delta_i$ with the derivative equal to $[F(\beta_i) - F(\alpha_i)]/(\beta_i - \alpha_i)$, but is also differentiable at each interior point $x \in P_{x_0}$. Indeed, $r(G, x, x + h) \rightarrow F'(x)$ as $h \rightarrow 0$ if $x + h \in P_{x_0}$; if, however, $x + h \in \sum_i \delta_i$, then it follows from the linearity of G on $\delta_i = (\alpha_i, \beta_i)$ that $r(G, x, x + h)$ is contained between two numbers $r(F, x, \alpha_i)$ and $r(F, x, \beta_i)$ whose limit as $h \rightarrow 0$ is $F'(x)$. If x is an end point of P_{x_0}, then in any case there exist at this point one-sided finite derivatives; thus the assertion that at each point there exists a finite right-hand derivative G'^+ is valid. In view of the absolute convergence of the series $\sum_i (F(\beta_i) - F(\alpha_i))$ and the summability of $F'(x)$ on P_{x_0}, the derivative G'^+ is summable on $[a, x_0]$; by Theorem 4.2', G is the integral of its derivative and

$$\int_a^{x_0} G'(x) \, dx = \sum_i \int_{\delta_i} G'(x) \, dx + \int_{Px_0} G'(x) \, dx$$

$$= \sum [G(\beta_i) - G(\alpha_i)] + \int_{P \cdot [a, x_0]} G'(x) \, dx,$$

i.e., Eq. (8.1) holds. (We see that the assumption that P is the set of points at which the function is nonsummable, is *not* used in the proof.)

---

[3] Let a number w_i correspond to each interval of some system $\{\delta_i\}$; the symbol $\sum_{\delta_i < x} w_i$ denotes the summation of those w_i that correspond to δ_i situated to the left of the point x. In the sum (8.1) the term of the form $F(x) - F(\alpha_i)$ is included if $x \in (\alpha_i, \beta_i)$.

8.2   DENJOY'S TOTALIZATION

We shall now discuss the content of Denjoy's two notes [1] and [2] mentioned earlier. The proofs to be presented are along the lines of the brief arguments by Denjoy which serve as justification for the theorems formulated in these notes. The extensive papers by Denjoy [4] and [5] contain a detailed analysis of totalization. The note by Lusin [3] (1912) is also devoted to this topic and the corresponding proofs are available in Lusin's thesis [4].

Let f be a given function. A *total* in Denjoy's sense is the number $V(a, b) = V(\delta)$ which depends on the interval (a, b) and which is calculated using the following definitions.

Definition 8.1

The total $V(a, b)$ in the interval (a, b) in which f is Lebesgue-summable is defined by

$$V(a, b) \stackrel{\text{def}}{=} \int_a^b f(x)\, dx.$$

Definition 8.2

If the total is defined on the intervals $(a_1, a_2), (a_2, a_3), \ldots, (a_{n-1}, a_n)$, then

$$V(a_1, a_n) \stackrel{\text{def}}{=} \sum_{i=2}^n V(a_{i-1}, a_i).$$

Definition 8.3

If f is summable on a perfect set P situated on (a, b), if the total is calculated on every subinterval δ_n' of CP, and if, moreover, the series $\sum w(\delta_n)$ converges (where $CP = \sum \delta_n$ and $w(\delta_n) = \sup_{\delta_n' \subset \delta_n} |V(\delta_n')|$), then

$$V(a, b) \stackrel{\text{def}}{=} \sum_n V(\delta_n) + \int_P f(x)\, dx. \tag{8.2}$$

The number $V(a, b)$ calculated by Definitions 8.1–8.3 is called the *restricted total* or the *restricted Denjoy integral* obtained as the result of *restricted totalization*.

(An additional Definition 8.2′ formulated below should be added to the set of Definitions 8.1–8.3.)

After formulating the definition of a restricted total, Denjoy states the conditions to be imposed on the function to assure its totalizability; these conditions should guarantee the applicability of the operations of Definitions 8.1–8.3.

Definition 8.4

A function f is called totalizable (in the restricted sense) on the interval (a, b) if it satisfies the following three conditions.

Condition I

For any perfect set P of $[a, b]$, the set of points of nonsummability of f on P is nowhere dense on P.

Condition 2 (Continuity of V)

If for any interval (a', b'), $a < a' < b' < b$, the total $V(a', b')$ is well defined, then the limit $\lim_{\substack{a' \to a \\ b' \to b}} V(a', b')$ exists. In this case *we define*

$$V(a, b) = \lim_{\substack{a' \to a \\ b' \to b}} V(a', b').$$

Remark. Condition 2 expresses a property of a restricted total which should have been included in the previous definitions. This property will be called *Definition 8.2'* in the following text. It asssures the continuity of the restricted total as a function of a and b.

Condition 3

For any perfect set P in the adjacent intervals of which the totals are calculated, the set of points of P for which the series $\sum w(\delta_n)$ is not convergent, is nowhere dense in P.

Following Denjoy, we will show how to compute the restricted total $V(a, b)$ of a function totalizable on (a, b).[4] Let $P_0 = [a, b]$. In view of

---

[4] For the construction of a total, operations are used whose meaning and applicability are described by Definitions 8.1–8.3. Other equivalent systems of definitions are also available. Variations in the definitions affect the sequence of operations carried out in the calculation of the total. For example, Natanson [1] and Saks [1] apply Definition 8.3 to closed sets. (For this reason there is need to pass from closed sets P to their perfect kernels utilizing Definitions 8.2 and 8.2'.)

Condition 1, there exists a set of portions everywhere dense on P_0 in which f is summable; these portions are all possible intervals situated inside the complement of the set P_1 of points of nonsummability of f on P_0. According to Definition 8.1, the totals are defined on all these portions. Hence, in accordance with Definition 8.2′, the totals are defined on all intervals adjacent to P_1 by Condition 2. Next, utilizing Definitions 8.2 and 8.2′, we define totals on the intervals adjacent to $(P_1)′$; the successive application of Definitions 8.2 and 8.2′ leads us to the definition of totals on intervals adjacent to any $(P_1)^{(\gamma)}$. Hence, we may assume that P_1 is perfect and nowhere dense on P_0.

Remark. Definitions 8.2 and 8.2′ are applied each time immediately following the application of Definition 8.3 in the course of the definition of the sets P_γ as well as in several other cases. The procedure resulting from the application of these definitions completely coincides with the procedure of constructing Di-integrals (Section 2.4). We shall omit this intermediate stage of constructions from our text in order not to clutter the exposition with arguments of the same kind.

We now continue with our arguments. Assume that for some finite or transfinite γ, perfect nonempty sets P_β have been defined for all $\beta < \gamma$ such that $P_{\beta'}$ is contained in P_β and is nowhere dense on it if $\beta' > \beta$, and let the totals be defined on intervals adjacent to each P_β. If γ is a number of the second kind, we shall assume as usual that $P_\gamma = \prod_{\beta < \gamma} P_\beta$. Clearly $P_\gamma \subset P_\beta$ for every $\beta < \gamma$ and hence, P_γ is nowhere dense on P_β.

We show now that the totals on every interval δ adjacent to P_γ may be calculated according to the rules (definitions) above. Indeed, every interval δ_1, which is located inside δ, belongs to the complement of some P_β, $\beta < \gamma$; but then in this case $V(\delta_1)$ is defined by the induction assumption. Hence, $V(\delta)$ is defined according to Definition 8.2′. Next we proceed as was stated in the remark above. In order not to introduce additional notation we shall assume that P_γ is a perfect set.

Now let γ be a number of the first kind. Then according to Condition 1, an everywhere dense set of portions on which f is summable can be found on $P_{\gamma-1}$; the totals are defined on the intervals adjacent to these portions according to the induction assumption. In view of Condition 3, a portion can be found in each one of these portions such that the series $\sum w(\delta_n)$ is convergent on this portion. Thus, $P_{\gamma-1}$ possesses an everywhere dense set of portions on which the total is defined according to

Definition 8.3. The complement of these portions relative to $P_{\gamma-1}$ is a nowhere dense set on $P_{\gamma-1}$ which, as we have seen above, can be assumed to be a perfect set. We denote this set by P_γ.[5]

Thus, by transfinite induction, a strictly decreasing sequence of perfect sets P_γ is defined on the adjacent intervals to which the total is calculated. By the Cantor–Baire stationarity principle, P_γ is empty for some γ and $CP_\gamma = (a, b)$. Thus, by means of a generally transfinite number of operations, the total in (a, b) is calculated as required.

We note that a priori $V(a, b)$ may depend not only on f but also on the method of construction, i.e., on the sequence $\{P_\gamma\}$. However, we shall show in what follows that there is actually no such dependence.

Finally, we must verify that $V(a, b)$ is indeed a total, i.e., that it satisfies the descriptive definition of a restricted total.[6] This verification is carried out at each step of the transfinite construction of the total.

Verification of the Conditions in Definitions 8.2 and 8.2' (finite additivity of totals, continuity of totals)

We show that a total possesses the properties (as given by Definitions 8.2 and 8.2') on any interval adjacent to P_γ for each γ. On those intervals where the function f is summable, the total is evidently additive and continuous (being a Lebesgue integral). In view of the remark on p. 138, it will be assumed proved that the total possesses these properties outside of P_1. Let it be assumed that on intervals adjacent to P_β, $\beta < \gamma$, the total is additive and continuous. It is directly verified that the total is continuous and additive also for $\beta = \gamma$, if γ is a number of the second kind. Now let γ be a number of the first kind and $\delta = (x', x'')$, $[x', x''] \subset CP_\gamma$, $C_\delta P_{\gamma-1} = \sum \tilde{\delta}_n$ and, furthermore, let the total on δ be defined by the formula

$$V(x', x'') = \sum V(\tilde{\delta}_n) + \int_{x'}^{x''} \varphi(x) \, dx, \tag{8.3}$$

where $\varphi(x) = \chi_{P_{\gamma-1}}(x) \cdot f(x)$ and $\chi_{P_{\gamma-1}}$ is the characteristic function of

---

[5] The choice of P_γ can be unique if the set of points of divergence of the series $\sum w(\delta_n)$ on $P_{\gamma-1}$ is considered; however, one should keep in mind that the totalization, in principle, is possible with different chains of the sets P_γ.

[6] Denjoy did not carry out this verification since he considered the collection of Definitions 8.1–8.3, not as a descriptive definition of a total, but as a set of rules for calculating totals for some functions. However, such a verification *is* necessary.

the set $P_{\gamma-1}$. Let $\delta^* = (x'', x''') \subset CP_\gamma$, $C_{\delta^*} P_{\gamma-1} = \sum_n \delta_n^*$ and a formula analogous to Eq. (8.3) be valid for $V(x'', x''')$. Clearly, $C_{(x', x''')} P_{\gamma-1} = \sum \tilde{\delta}_n + \sum \delta_n^*$ if $x'' \in P_{\gamma-1}$, and $C_{(x', x''')} P_{\gamma-1} = \sum \tilde{\delta}_n + \sum \delta_n^* + \{x''\}$ if x'' belongs to the interval (s, t) adjacent to $P_{\gamma-1}$. In the first case the equality

$$V(x', x''') = \sum V(\tilde{\delta}_n) + \sum V(\delta_n^*) + \int_{x'}^{x''} \varphi(x)\, dx + \int_{x''}^{x'''} \varphi(x)\, dx$$

$$= V(x', x'') + V(x'', x''')$$

follows immediately. The same equality holds also in the second case if one observes that by the induction hypothesis, the additivity is valid outside of $P_{\gamma-1}$ and hence,

$$V(s, x'') + V(x'', t) = V(s, t).$$

We now prove the continuity of the right-hand side of (8.3) as a function of x', x''.[7] It is sufficient to prove the continuity of the sum in the right-hand side of (8.3) *as a function* of $x(x \equiv x'')$ at the points of the set $P_{\gamma-1}$ belonging to an interval δ adjacent to P_γ, $C_\delta P_{\gamma-1} = \sum \delta_n$ (in $CP_{\gamma-1}$, the continuity holds by the induction assumption) provided that the series $\sum w(\delta_n)$ is convergent. Denote this function by θ. Define for a given ε a number N such that $\sum_{n \geqslant N} w(\delta_n) < \varepsilon$. Let $|h| < \min(\delta_1, \ldots, \delta_N)$; then it follows from $\delta_n(x, x + h) \neq 0$ and $x + h \in P_{\gamma-1}$ that $n > N$ and

$$|\theta(x + h_1) - \theta(x)| \leqslant \sum_{\delta_n \subset (x, x+h)} w(\delta_n) < \varepsilon,$$

provided $|h_1| < |h|$.

If x is an end point of $P_{\gamma-1}$, then the continuity from (the side of) the adjacent interval follows from Definition 8.2′ and the induction assumption.

---

[7] One may contend that no proof is needed since a total is continuous by Condition 2 of totalizability (p. 137). We note, however, that this condition is actually utilized only when eliminating isolated points in the derived sets P_β and in constructing the total outside of P_γ in the case when γ is of the second kind; if Condition 2 is satisfied in these two cases *only*, the continuity of the restricted total follows (everywhere).

Verification of the Condition in Definition 8.3

Let P be a perfect set on which f is summable, $CP = \sum \delta_n$, and let the series $\sum w(\delta_n)$ be convergent; we are required to show that Eq. (8.2) is valid. We shall prove that Eq. (8.2) is valid on the portions of P situated inside CP_γ.

Indeed, if $\delta \cdot P$ is a portion inside CP_1 and f is summable on δ, then

$$V(\delta) = \int_\delta f(x)\, dx = \int_{\delta \cdot P} f(x)\, dx + \sum_{\delta_i \subset \delta} \int_{\delta_i} f(x)\, dx$$

$$= \int_{\delta \cdot P} f(x)\, dx + \sum_{\delta_i \subset \delta} V(\delta_i).$$

In view of the remark on p. 138, we shall extend the last equality over arbitrary intervals δ, situated on CP_1. Thus, for the portions $\delta \cdot P$ in CP_1, Eq. (8.2) is valid. Assume that it is valid for portions of P, located in CP_β, $\beta < \gamma$; it follows from this assumption that it is valid for portions of P in CP if γ is of the second kind. Let γ be of the first kind, let $\delta \cdot P$ be a portion of P inside CP_γ, $\delta \cdot CP_{\gamma-1} = \sum \tilde{\delta}_j$ and let the total $V(\delta')$ in every interval δ', located in δ, be defined by the formula

$$V(\delta' \cdot \delta) = \int_{\delta' \cdot P_{\gamma-1}} f(x)\, dx + \sum_j V(\delta' \cdot \tilde{\delta}_j); \qquad (8.3')$$

by the induction assumption,

$$V(\tilde{\delta}_j) = \int_{\tilde{\delta}_j \cdot P} f\, dx + \sum_i V(\delta_i \cdot \tilde{\delta}_j).$$

Putting $\delta' = \delta$ in (8.3') and substituting the expression obtained for $V(\tilde{\delta}_j)$, we obtain

$$V(\delta) = \int_{\tilde{\delta} \cdot P_{\gamma-1}} f\, dx + \sum_j \int_{\tilde{\delta}_j \cdot P} f\, dx + \sum_j \sum_i V(\delta_i \cdot \tilde{\delta}_j); \qquad (8.3'')$$

since $\delta_i \cdot \delta \subset \delta$, we also have

$$V(\delta_i \cdot \delta) = \int_{\delta \cdot \delta_i \cdot P_{\gamma-1}} f\, dx + \sum_j V(\delta_i \cdot \tilde{\delta}_j); \qquad (8.3''')$$

rewriting (8.3'') in the form

$$V(\delta) = \int_{\delta \cdot P \cdot P_{\gamma-1}} f\, dx + \sum_i \int_{\delta_i \cdot P} f\, dx$$

$$+ \sum_i \left(\int_{\delta_i \cdot P_{\gamma-1}} f\, dx + \sum_j V(\delta_i \cdot \tilde{\delta}_j) \right)$$

and utilizing (8.3'''), we finally obtain

$$V(\delta) = \int_{\delta \cdot P} f \, dx + \sum_i V(\delta_i \cdot \delta). \qquad \text{Q.E.D.}$$

We now discuss the second note in Volume 154 of *Comptes Rendus*. In this note Denjoy [2] formulates and gives partial proofs of the main theorems of totalization.

Theorem 8.1

The restricted total $V(a, x)$ is a continuous function of x.

This theorem was proved in the course of the verification of Definition 8.2'.

Theorem 8.2

The derivative of an indefinite restricted total exists almost everywhere and equals the totalized function almost everywhere.

To prove the theorem, we use the following auxiliary assertion which was proved by Denjoy.

Assertion 1

If P is a perfect nowhere dense set on $[a, b]$, $CP = \sum \delta_n$, $\sum a_n$ is an absolutely convergent series, and

$$\theta_1(x) \overset{\text{def}}{=} \sum_{\delta_n < x} a_n,$$

then $\theta_1'(x) = 0$ almost everywhere.

Proof

Clearly, it is sufficient to consider the case of positive a_n; then $\theta_1(x)$, being a monotonically increasing function, possesses a derivative $\theta_1'(x)$ almost everywhere. Assume that $\theta_1'(x) > \alpha > 0$ on the set E_α and let

$$E_n \overset{\text{def}}{=} E_x\left(x \in E_\alpha, \ \frac{\theta_1(x + h) - \theta_1(x)}{h} > \alpha, \ |h| \leqslant \frac{1}{n} \right).$$

We have $E_n \subset E_{n+1}$ and $\lim_{n \to \infty} E_n = E_\alpha$. Consider a partition $\sigma = \{x_i\}$ of the segment $[a, b]$ with $d(\sigma) < 1/n$, and let Δ_i be those partition segments which contain points of E_n; it follows from the definition of E_n and the condition $d(\sigma) < 1/n$ that $\theta_1(\Delta_i)/m\Delta_i > \alpha$; summing up over all such Δ_i, we obtain that $\theta_1([a, b]) \geqslant \sum_i \theta_1(\Delta_i) \geqslant \alpha \sum m\Delta_i$; since $\sum \Delta_i \supset E_n$, we have $\theta_1([a, b]) \geqslant \alpha m E_n$, and we finally obtain

$$\theta_1([a, b]) \geqslant \alpha m E_\alpha \tag{8.4}$$

as $n \to \infty$.

We now note that the set E is not changed if, in place of θ_1, the function

$$\theta_k(x) \stackrel{\text{def}}{=} \sum_{\delta_n < x} \tilde{a}_n$$

is considered, where $\tilde{a}_n = 0$ for $n < k$ and $\tilde{a}_n = a_n$ for other values of n. Indeed, as long as the length of the segment Δ is less than the length of each one of δ_i, $i = 1, 2, \ldots, k - 1$, the inclusion $\delta_i \subset \Delta$ is not valid for these i, and if additionally, it is assumed that the end points of Δ belong to P, then $\theta_1(\Delta) = \theta_k(\Delta)$; hence, θ_1 and θ_k are both differentiable and $\theta_1'(x) = \theta_k'(x)$. Thus, we obtain from (8.4) the formula

$$\theta_k([a, b]) \geqslant \alpha m E_\alpha \tag{8.4'}$$

valid for any k. Since $\lim_{k \to \infty} \theta_k(x) = 0$, $m E_\alpha = 0$; but α is arbitrary, hence, $\theta_1'(x) = 0$ almost everywhere. Assertion 1 is thus proved.

Remark (Khinchin [2]). In particular, it follows from Assertion 1 that for almost all points of density x of the set P, the relation

$$\lim_{d \to 0} \frac{a_n}{d} = 0 \tag{8.5}$$

is satisfied, where d is the distance from the point x to the nearest end point of the interval δ_n.[8]

---

[8] The meaning of the left-hand side of (8.5) is as follows: x is fixed, and $d \to 0$ running through the values equal to the distance from point x to the nearest end points of the intervals δ_n; for each such d there corresponds an n.

Indeed, let d^* be the distance from the point x to the furthest end point of δ_n; then it follows from the definition of a density point that

$$\lim_{d \to 0} \frac{d}{d^*} = 1;$$

but if x is a point such that $\theta_1'(x) = 0$, then, certainly,

$$\lim_{d^* \to 0} \frac{a_n}{d^*} = 0,$$

and (8.5) follows from the last equation.

Proof of Theorem 8.2

Arguments analogous to those given in the proof of the continuity of $V(a, x)$ yield that it is sufficient to prove the differentiability with respect to $x'' = x$ of the right-hand side of Eq. (8.3) at almost all points of the set $P_{\gamma-1}$ belonging to the interval δ. The integral in this formula is differentiable with respect to the upper limit almost everywhere and, moreover, its derivative is equal to f everywhere on $\delta \cdot P_{\gamma-1}$. Hence, it is sufficient to prove that $\theta'(x) = 0$ almost everywhere on $\delta \cdot P_{\gamma-1}$ where θ is the sum in the right-hand side of (8.3). Let $C_\delta P_{\gamma-1} = \sum \delta_n$; put $a_n = w(\delta_n)$ in Assertion 1; let $x_0 \in \delta \cdot P_{\gamma-1}$ be a point at which $\theta_1'(x_0) = 0$ and at which relation (8.5) is satisfied. We show that $\theta'(x)_0 = 0$. Indeed, if $x_0 + h \in P_{\gamma-1}$, then $\theta(x_0 + h) - \theta(x_0) = \theta_1(x_0 + h) - \theta_1(x_0)$; if, however, $x_0 + h \in \tilde{\delta}_k = (\alpha_k, \beta_k)$, then

$$\frac{|\theta(x_0 + h) - \theta(x_0)|}{|h|} \leqslant \frac{|\theta_1(\alpha_k) - \theta_1(x_0)|}{|h|} + \frac{w(\tilde{\delta}_n)}{|h|}.$$

Under the above assumptions on x_0, the right-hand side of the last inequality tends to zero as $h \to 0$. Theorem 8.2 is thus proved.

Theorem 8.3

An exact finite derivative is totalizable in the restricted sense.

Proof

Let F be a function differentiable at each point $x \in [a, b]$, $F'(x) = f(x)$. We show that f satisfies the three conditions of totalizability.

Verification of Condition 1

f is a function of *Baire's class one* (Section 4.7). On the basis of the Baire theorem on functions of class one, we conclude that on every perfect set P there is an everywhere dense set of continuity points of f (with respect to P). For each such point there exists a neighborhood in P in which f is bounded and, hence, summable. The set of these neighborhoods is everywhere dense on P.   Q.E.D.

Verification of Condition 2

If it is assumed that the equality $V(a', b') = F(b') - F(a')$ holds, the validity of the condition follows from the continuity of F. This assumption is justified since the equality in question holds inside the adjacent intervals to the sets P_γ. These equalities in turn are verified at each step of the transfinite construction.[9]

Verification of Condition 3

Let P be a perfect set, $CP = \sum \delta_i$. Assume that the equality $V(\delta) = F(\delta)$ is satisfied on the intervals δ_i. As we noted above, the totality of portions of the set P, on which f if bounded, is everywhere dense on P. Let $\delta_1 \cdot P$ be an arbitrary portion P on which $|f(x)| < M$; from the equality

$$\delta_1 \cdot P = \sum_n \mathop{E}_{x \in \delta_1 \cdot P} \left(|r(F, x, x + h)| \leqslant M, \ |h| \leqslant \frac{1}{n} \right) = \sum_n E_n,$$

the Baire category theorem, and the fact that E_n is closed, the existence of a portion $\delta_2 \cdot P, \delta_2 \cdot P \subset \delta_1 \cdot P$, belonging to some E_n, follows. Take a portion $\delta P \subset \delta_2 P$ of diameter smaller than $1/n$. Then

$$|r(F, x, x + h)| \leqslant M, \qquad x \in \delta \cdot P, \quad x + h \in \delta. \qquad (8.6)$$

In the final analysis, there exists a set G of portions δP, everywhere dense on P, in each one of which condition (8.6) is satisfied for some M.

We show that the series $\sum_{\delta_i \subset \delta} w(\delta_i)$ is convergent. Note that

$$w(\delta_i) = \sup_{x_1, x_2 \in \delta_i} |F(x_2) - F(x_1)|.$$

---

[9] This formal Condition 2 is necessary since it is used for construction of a specific chain P_γ. The possibility of constructing such a chain yields the fact that the condition is satisfied in all cases. Similar remarks are valid for Condition 3. Therefore, it is sufficient to verify here whether the construction of the chain $\{P_\gamma\}$ is possible.

Denote $\delta_i = (\alpha_i, \beta_i)$; since $P \cdot \delta_i \neq 0$, let, for example, $\alpha_i \in \delta \cdot P$. Then

$$|r(F, \alpha_i, x_j)| < M, \qquad j = 1, 2.$$

From here we immediately obtain

$$|F(x_1) - F(x_2)| < 2Mm\delta_i,$$

which proves the convergence of the series $\sum_{\delta_i \subset \delta} w(\delta_i)$.

Thus the set of points of divergence of this series is situated outside of the portions $\delta \cdot P$, i.e., this set is nowhere dense on P. Theorem 8.3 is thus proved.

Theorem 8.4

The indefinite restricted total $V(a, x)$ of an exact finite derivative is a primitive function.

Proof

Let $F' = f$ on $[a, b]$. In the portions situated inside CP_1, the total being the Lebesgue integral of the function f, is equal to the increment of the derivative.

Three cases may arise in the construction of totals:

(1)   The total is constructed using Definition 8.2. Then

$$\begin{aligned}
V(a_1, a_n) &= V(a_1, a_2) + \cdots + V(a_{n-1}, a_n) \\
&= F(a_2) - F(a_1) + \cdots + F(a_n) - F(a_{n-1}) \\
&= F(a_n) - F(a_1).
\end{aligned}$$

(2)   The total is constructed using Definition 8.2'. Then the equality

$$V(a', b') = F(b') - F(a')$$

remains valid as $a' \to a$, $b' \to b$, in view of the continuity of the total and of the function F.

(3)   The total $V(\delta)$ is obtained utilizing Definition 8.3. In this case it coincides with the increment of the derivative: this follows from the theorem proved in Section 8.1. Theorem 8.4 is thus proved.

In concluding his note Denjoy expresses his belief that it is possible to totalize finite Dini derivatives; he notes that in this case the last two conditions of totalizability are satisfied (the second with $w(\delta_n)$ replaced by $V(\delta_n)$!) It seems that at that time, Denjoy could not prove that the first condition is also satisfied. Totalization of derivatives is discussed in detail in Denjoy's memoirs [4], [5].

Remark. The condition of convergence of the series $\sum w(\delta_n)$ plays a special role in Definition 8.3. A careful reader may have noticed that this condition is not necessary for the construction of a total; the series $\sum V(\delta_n)$ appears in these constructions and its absolute convergence does not imply the convergence of the first series. However, the convergence of the series $\sum w(\delta_n)$ is used substantially in the proof of differentiability of restricted totals. It was unknown at that point whether this condition was necessary for a restricted total to be differentiable. Later Khinchin [1], [2] showed that this condition can be relaxed (see Section 8.3 below).

In view of this remark the following definition is in order:

Definition 8.5

The number $V(a, b)$ is called the general total or the general Denjoy integral[10] if it is obtained using Definitions 8.1, 8.2, 8.2', and 8.3, while in Definition 8.3 the condition of convergence of the series $\sum_n w(\delta_n)$ is replaced by the condition of absolute convergence of the series $\sum_n V(\delta_n)$.

There is no doubt that the feasibility of such a totalization was known to Denjoy and others from the very beginning (see the remark above by Denjoy on the possibility of totalization of derivatives); however, the formulation of this fact was given only in 1916 simultaneously by Khinchin and Denjoy.

Denjoy arrived at the definition of a restricted total during his search for a procedure leading to a primitive function; next, this problem naturally motivated the investigation of totalization of functions given in advance. Additional investigations by Lusin, Denjoy, and Khinchin led to a descriptive definition of Denjoy integrals.

---

[10] This term is used by Lebesgue. Natanson [1] and Saks [1], use the term "Denjoy integral in the wide sense."

8.3   A DESCRIPTIVE DEFINITION OF DENJOY INTEGRALS. KHINCHIN'S INTEGRAL

Denjoy's notes in *Comptes Rendus* were noticed by Lusin at once, and he published the first descriptive definition of an indefinite restricted Denjoy integral in the following volume (*Comptes Rendus* **155**). Later in 1916 Denjoy formulated a descriptive definition of an integral obtained by general totalization, i.e., of a general Denjoy integral; Khinchin [1] in 1916 replaced the condition of convergence of the series $\sum w(\delta_n)$ by a condition which is necessary and sufficient for the totalizable function to be almost everywhere the derivative of the indefinite total. He also showed simultaneously with Denjoy that in the case of general totalization, the totalized function is an approximate (asymptotic) derivative of its indefinite total.

*What are the basic properties of a general indefinite total $F(x) = V(a, x)$ which may serve as its definition?

(1)   It follows from Condition 2 of totalizability that an indefinite total F is a continuous function. (The continuity of the *general* total is actually postulated by Condition 2.)

(2)   Conditions 1 and 3 and Definition 8.3 [with the condition of convergence of the series $\sum w(\delta_n)$ replaced by the condition of absolute convergence of the series $\sum V(\delta_n)$] imply that on every perfect set P a portion $\delta \cdot P$ can be found $(C_\delta P = \sum \delta_n)$ such that F is expressed on this portion (up to a constant) using Eq. (8.2). This property can also be expressed in the following manner [cf. the proof of the last assertion in Eq. (8.1)]: on every perfect set P, a portion $\delta \cdot P$ can be found such that a continuous function G, which coincides on $\delta \cdot P$ with F and which is linear on the intervals adjacent to $\delta \cdot P$, is absolutely continuous. However, how can one formulate this property in terms of the function F considered only on P? We thus arrive at a generalization of absolute continuity for the case when an interval is replaced by an arbitrary set. This is given in the next definition.*

Definition 8.6   (Khinchin [1])

A function F is called absolutely continuous on the set E if for every $\varepsilon > 0$, a $\delta > 0$ can be found such that

$$\sum_i |F(b_i) - F(a_i)| < \varepsilon, \tag{8.7}$$

where $\{(a_i, b_i)\}$ *is a sequence of nonoverlapping intervals whose end points belong to E and* $\sum(b_i - a_i) < \delta$.

*We now show that the condition of absolute continuity of the function G on δ is equivalent to the condition of absolute continuity of F on $\delta \cdot P$.

Indeed, let F be absolutely continuous on $\delta \cdot P$. If in formula (8.7) in place of (a_i, b_i) we use intervals δ_i adjacent to $\delta \cdot P$ (where i is sufficiently large), then the series $\sum G(\delta_i) = \sum F(\delta_i)$ is absolutely convergent. The increment of the function G on any segment belonging to δ does not exceed the sum of three increments: two increments on segments belonging to the segments δ_i and an increment on a segment with end points in $\delta \cdot P$. The sum of the increments on the segments belonging to δ_i is bounded by the remainder of the series $\sum |G(\delta_i)|$ (G is linear on each δ_i). Since G is absolutely continuous on $\delta \cdot P$, the sum of the increments on segments with end points in $\delta \cdot P$ is arbitrarily small. Hence, we have shown that G is absolutely continuous on δ.

We can therefore express the second property of a total in the following manner: *for any perfect set P, the function F is absolutely continuous on some of its portions $\delta \cdot P$.*

What remains is to establish the connection between the total and the totalizable function. In the case of restricted totalization, the totalizable function is the derivative almost everywhere of the indefinite total (Theorem 8.2). In the case of general totalization, the totalizable function is an "approximate derivative" almost everywhere of the indefinite total.*

Definition 8.7 (Denjoy [3], Khinchin [1])

The function F is called approximately derivable at x if, for some set P with a point of density at x, there exists the limit

$$\lim_{h \to 0} \frac{F(x + h) - F(x)}{h}$$

as $x + h$ approaches x along P.

It follows directly from Definition 8.7 that a function cannot have several approximate derivatives at the point x; clearly a function which is differentiable in the usual sense is approximately derivable (differentiable). (More details on approximate differentiability of functions are given by Saks [1].)

The Denjoy–Khinchin Theorem (Denjoy [4],[5]; Khinchin [1],[2])

The indefinite general total F of a totalizable function f possesses almost everywhere an approximate derivative equal to f.

Proof

F is an indefinite Lebesgue integral of f, inside the intervals adjacent to the set of points of nonsummability of the function, and hence, f is the ordinary derivative of F almost everywhere inside these intervals. Assume that the assertion of the theorem is valid at the intervals adjacent to P_β, $\beta < \gamma$. If γ is of the second kind, then clearly the theorem is valid also in the intervals adjacent to P_γ. If γ is of the first kind, we assume that in the interval δ situated inside CP_γ, the total is determined by the Eq. (8.3). Then if x, $x + h \in \delta \cdot P_{\gamma-1}$, we have

$$\frac{F(x + h) - F(x)}{h} = \frac{1}{h} \int_{P_{\gamma-1} \cdot [x, x+h]} f(x)\, dx + \frac{1}{h} \sum_{\delta_n \subset [x, x+h]} V(\delta_n),$$

where $\sum \delta_n = C_\delta P_{\gamma-1}$. Putting in Assertion 1 (p. 142), $a_n = V(\delta_n)$, we obtain

$$\lim_{h \to 0} \frac{1}{h} \sum_{\delta_n \subset [x, x+h]} V(\delta_n) = 0$$

at almost all points x in $\delta \cdot P_{\gamma-1}$. Thus almost everywhere in $\delta \cdot P_{\gamma-1}$, the limit

$$\lim_{\substack{h \to 0 \\ x+h \in P_{\gamma-1}}} \frac{F(x + h) - F(x)}{h} = f(x)$$

exists. This limit is the approximate derivative if x is a point of density of $P_{\gamma-1}$. The theorem is thus proved.

We are now in a position to formulate the descriptive definition of a total obtained by general totalization (see Remark 1 below).

Definition 8.8

A function F is called an indefinite general total of the function f in [a, b] if:

(1) *F is continuous;*

(2) *F possesses an approximate derivative equal to f almost everywhere on [a, b];*

(3) *for any perfect set $P \subset [a, b]$, F is absolutely continuous on some portion of P.*

A continuous function *F* which possesses property (3) will be called, following Denjoy, a *resolvable (résoluble)* function. Clearly, every resolvable function is approximately derivable almost everywhere and is obtained from its approximate derivative by 'the process of general totalization.

Condition (3) can be replaced by the following condition: it is possible to represent the segment [a, b] as an at most countable sum of sets E_i, $[a, b] = \sum_i E_i$ and on each E_i the function f is absolutely continuous (Khinchin [1] expresses Condition (3) in this form). Indeed, the sets E_i can be constructed from the portions of the sets P_β belonging to the adjacent intervals of the set $P_{\beta+1}$ on which F is absolutely continuous for all β. Conversely, if the above-stated decomposition of the segment [a, b] is possible, then for a perfect set $P \subset [a, b]$ we have $P = \sum_i P \cdot E_i$. According to the Baire category theorem, some set $P \cdot E_j$ is dense on some portion $\delta \cdot P$. But if a continuous function is absolutely continuous on a dense subset, then it is also absolutely continuous on the whole set. Hence, F is absolutely continuous on $\delta \cdot P$ and Condition (3) is satisfied.

Finally, we show that Conditions (1)–(3) *completely determine* the function F, i.e. there are no two distinct resolvable functions with approximate derivatives equal almost everywhere. Indeed, let F be resolvable and let the approximate derivative $F'_{ap} = 0$ almost everywhere. Let P be the set of points in every neighborhood of which F is a nonconstant function. A portion $\delta \cdot P$ can be found on which F is absolutely continuous, and in view of the definition of P, F is absolutely continuous on δ. Thus,

$$F(\delta_1) = \int_{\delta_1} F'_{ap} \, dx = 0$$

in any portion $\delta_1 \subset \delta$ and hence, P is empty.

It follows from the theorem proved above that the result of totalization does not depend on the chain $\{P_\gamma\}$ (see p. 139).

Remark 1. Definition 8.8 was given in a somewhat different form by Denjoy [3] and was analyzed in detail in his papers [4] and [5]. Denjoy's equivalent formulation of resolvability is as follows: a continuous function F is resolvable if every perfect set P of measure zero contains a portion $\delta \cdot P$, $C_\delta P = \sum \delta_i$ such that the series consisting of increments in the adjacent intervals δ_i is absolutely convergent and also $F(\delta) = \sum F(\delta_i)$. The equivalence of this and the preceding definitions of resolvability is a fine point in the theory of functions. The proof of this fact is a corollary of a number of assertions presented by Denjoy in [4]; it is also available in Lebesgue [4] and [5].

Remark 2. Khinchin [1], along with Denjoy [3], formulated the theorem on approximate derivability of a general total (the proof was given in Khinchin [2]), interpreting it as the result of totalization of a certain function. He also formulated a descriptive definition of the general total (in the form of a theorem characterizing Khinchin's integral, see below). In these papers, Khinchin dealt mainly with a certain intermediate type of totalization which will be discussed in Section 8.5.

Denjoy [3], [4] has shown that a continuous function which possesses an approximate derivative at each point is a general total. In these papers Denjoy also proved the resolvability of a number of important classes of functions.

We stress the complete analogy of the following notions: an indefinite general total (the result of general totalization), resolvable functions, and an approximate derivative on one hand, and an indefinite Lebesgue integral, absolutely continuous functions, and an ordinary derivative on the other. Indeed,

(a)   every absolutely continuous function is an indefinite Lebesgue integral of its derivative;

(b)   every resolvable function is an indefinite general total of its approximate derivative.

(For more detailed treatment of problems connected with Denjoy's totalization we refer the reader, in addition to the original treatises of Denjoy and Lusin, to Saks [1] and Leçons II of Lebesgue [4].)

8.4   A DESCRIPTIVE DEFINITION OF THE RESTRICTED
DENJOY INTEGRAL

Returning to the analysis of the notion of total, we note that in a case of restricted totalization the following condition should supplement property (2) of absolute continuity of the general total F on a portion $\delta \cdot P$ (see p. 149):

(2′)   The series

$$\sum \omega(F, \delta_n), \tag{8.8}$$

with $\sum \delta_n = C_\delta P$, is convergent.

Property (2) imposes a restriction on the behavior of F *on the set* $\delta \cdot P$. Condition (2′), together with condition (2), imposes a restriction on the behavior of F *in an arbitrarily small neighborhood of the set* $\delta \cdot P$; we thus arrive at the following definition.

Definition 8.9

The function F defined on $[a, b]$ is called absolutely continuous in the restricted sense on a perfect set P, $P \subset [a, b]$ if, for every $\varepsilon > 0$, there exists a $\delta > 0$ such that for any system of nonoverlapping intervals $\{\delta_i\}$, each one of which containing a point of P such that $\sum_i \delta_i \supset P$ and $\sum_i m\delta_i < \delta$, one has $\sum_i \omega(F, \delta_i) < \varepsilon$.

*Clearly, a function that is absolutely continuous in the restricted sense on E is *a fortiori* absolutely continuous on E in the sense of Definition 8.6. Moreover, the set of conditions (2) and (2′) is equivalent to the following condition: *for any perfect set P there exists a portion of the set P on which F is absolutely continuous in the restricted sense.* The proof is similar to the proof of the equivalence of condition (2) and absolute continuity of G of δ (see p. 149).*

Definition 8.10

A function F is called an indefinite restricted total of the function f on (a, b) if:

(1)   *F is continuous on $[a, b]$;*
(2)   *F possesses almost everywhere a derivative equal to f;*
(3)   *for any perfect set P there exists a portion of P on which F is absolutely continuous in the restricted sense.*

This is essentially Lusin's [3] definition; it was formulated by Lusin as a theorem stating that the class of continuous functions possessing property (3) coincides with the class of functions obtained by restricted totalization.

Remark. Lusin introduced the notion of *variation of a continuous function on a perfect set P* as the limit of the sums of the form $\sum |F(\delta_i)|$, where $\{\delta_i\}$ is an arbitrary system of nonoverlapping intervals covering P such that each of them contains points of P, and the limit is taken as $\max_i m\delta_i \to 0$. According to Lusin, the function F is called a *function of generalized bounded variation on* $[a, b]$ if every perfect set $P \subset [a, b]$ contains a portion on which the variation of F is finite. In place of condition (3) the following equivalent condition is proposed by Lusin: the function F must be a function of generalized bounded variation and its variation on every set of measure zero must equal zero (if it exists).

It can be shown that as in the case of the indefinite general total, *a function which satisfies conditions* (1) *and* (3) *of Definition 8.10 is differentiable almost everywhere and is obtained from its derivative by the process of restricted totalization. Moreover, it is determined by its derivative* (*uniquely up to a constant*); *hence, the result of restricted totalization does not depend on the chain* $\{P_\gamma\}$.

Here again an analogy is valid for the notions of an indefinite restricted total (obtained by the process of restricted totalization), continuous functions satisfying condition (3) given in Definition 8.10, and an ordinary derivative on one hand, and on the other, an indefinite Lebesgue integral, absolutely continuous functions, and again, an ordinary derivative.

8.5   KHINCHIN'S INVESTIGATIONS

We have already referred to Khinchin's note [1] published in 1916 (with detailed proofs in [2]). On several occasions in this note, Khinchin points out the possibility of general totalization and formulates a theorem on the approximate differentiability of the general total. However, the main emphasis in this note is on a necessary and sufficient condition for the totalized function to be the derivative almost everywhere of its total. In the remark, on p. 147, we point out that the condition of convergence of the series $\sum w(\delta_n)$ is sufficient for differentiability almost everywhere of the indefinite total. Khinchin has shown that this condition can be replaced by a less restrictive one which is also necessary. Khinchin introduced the following definition:

Definition 8.11

We say that at the point x_0 belonging to a perfect set P, $CP = \sum \delta_n$, the totals are asymptotically annihilated if $w_n/d_n \to 0$, where w_n is the oscillation of the total in δ_n and d_n is the distance from the point x_0 to the nearest end point of δ_n.

Khinchin's Theorem

In order that the operation of general totalization lead to a total with a derivative almost everywhere, which is equal to the totalized function, it is necessary and sufficient that every perfect set P contain a portion such that at almost all points of the portion the totals are annihilated [Condition (0) *in our terminology*].

The *sufficiency* of the condition of the theorem follows directly by observing that in the proof of Theorem 8.2 we utilized only the fact that at the point x_0, at which the differentiability was proved, the totals were asymptotically annihilated [Eq. (8.5)].

The *necessity* in Khinchin's theorem should be interpreted as follows: if the total is differentiable almost everywhere then Condition (0) is satisfied. We show that Condition (0) is satisfied at each point of density $x_0 \in P$ at which F is differentiable. Write

$$F(x_0 + h_i) - F(x_0) = F'(x_0)h_i + \varepsilon_i h_i, \qquad i = 1, 2.$$

If the points $x_0 + h_1$, $x_0 + h_2$ belong to a single adjacent interval δ_n, then it follows from the last equality that

$$\omega(F, \delta_n) \leqslant 2\,|F'(x_0)|\,m\delta_n + 2\,|\varepsilon_1 h_1| + 2\,|\varepsilon_2 h_2|. \tag{8.9}$$

But $m\delta_n/d_n \to 0$ almost everywhere on P (see the remark on p. 143). Dividing both sides of the inequality (8.9) by $m\delta_n$ and taking the limit we find that the totals are asymptotically annihilated at the point x_0.

Clearly, a new type of totalization is defined on the basis of Khinchin's theorem. This is *totalization in Khinchin's sense* in which Definition 8.2 is used with the additional condition that almost everywhere on P, the totals be asymptotically annihilated (without requiring the convergence of the series $\sum w(\delta_n)$). The next theorem, due to Khinchin, contains a descriptive definition of Khinchin's total.

Theorem (Khinchin [1])

In order that a continuous function F, differentiable almost everywhere on the segment $[a, b]$, be an indefinite total in Khinchin's sense of its derivative F' on this segment, it is necessary and sufficient that the segment $[a, b]$ be decomposed into a finite or a countable sum of sets on which F is absolutely continuous.

The necessary and sufficient condition stated in this theorem is equivalent to the requirement that F' lend itself to general totalization; thus Khinchin was aware of the descriptive definition of general totals although it was not formulated explicitly in his paper [1].

Denjoy's integral satisfies Conditions (1)–(6) of the integration problem (Chapter 4). Conditions (1)–(5) are more or less obvious; as far as Condition (6) is concerned, it is sufficient to note that a totalizable nonnegative function is Lebesgue-integrable. If f_n are increasing monotonically and tend to f, then the difference $f - f_n$ is nonnegative, and to obtain the required conclusion it is sufficient to apply the theorem on term-by-term integration in the Lebesgue sense.

Concluding our exposition of the theory of Denjoy's integral, we repeat the opinion also expressed by other authors that Denjoy's integral is a "lucky" combination of Dirichlet's and Harnack's ideas implemented for the case of the integral in the Lebesgue sense. Indeed, Dirichlet's idea is used in each case with the application of Definitions 8.2 and 8.2' for the construction of the total. These definitions allow us to replace, for example, a closed set P_β by a perfect set P_β. Harnack's idea is clearly utilized when we apply Definition 8.3 and Eq. (8.2): the latter is merely an extension of Moore's [see Eq. (2.5), p. 21] and Hahn's formulas.

8.6 INTERRELATIONS BETWEEN DENJOY'S INTEGRAL AND OTHER INTEGRALS

If f is summable on $[a, b]$, then P_1 is empty and the totalization is completed in this case with one operation, the Lebesgue integration over $[a, b]$. Hence, if the function f is summable, then its Denjoy integral is equivalent to Lebesgue's integral. What is, however, the relation between various forms of totalization? The situation here is as follows: *totalization in the Khinchin sense is substantially more general than restricted totalization; general totalization is substantially more general than totalization in the Khinchin sense.*

To prove this, consider, for example, a perfect nowhere dense set $P \subset [0, 1]$, $CP = \sum \delta_i$, and a continuous function F equal to zero on P, differentiable on each δ_i and such that the series $\sum_i \omega(F, \delta_i)$ diverges at every point of the set P. F is evidently not a restricted Denjoy integral (total); however, it is easily seen that F is resolvable and, hence, is a general Denjoy integral. If $mP = 0$, then F' exists almost everywhere and F is a total in the Khinchin sense. Finally, to assure the nondifferentiability of F on a set of a *positive* measure we require that $mP > 0$; take a perfect subset P_1 of the set P $(mP_1 > 0)$, which does not contain the end points of P, and define F on δ_i in such a manner that $d(P_1 \cdot \delta_i) = o(\omega(F, \delta_i))$. Then indeed F will be nondifferentiable on P_1, and thus will not be a total in the Khinchin sense.

Consider now conditional methods of integration, namely, the Dirichlet–Hölder method and the Harnack method, and replace the Riemann integrals appearing in the definitions of the corresponding integrals by Lebesgue integrals (see, e.g., Lebesgue [13] and Lusin [4]). We thus consider Dirichlet–Lebesgue and Harnack–Lebesgue integrals. The corresponding definitions follow.

Definition of Integrability in the Dirichlet–Lebesgue Sense

A function f is called integrable if there exists a unique continuous function F satisfying on every interval (α, β) of summability of f the relation

$$F(\beta) - F(\alpha) = (\mathrm{L}) \int_\alpha^\beta f(x)\, dx.$$

Definition of Integrability in the Harnack–Lebesgue Sense

A function f is called integrable if the set N of its points of non-summability is discrete (or equivalently of Lebesgue measure zero, since N is closed) *and there exists a limit of the integral* (L) $\int_{C \sum \Delta_i} f(x)\ dx$ *as* $m \sum \Delta_i \to 0$; $\{\Delta_i\}$ *is a finite system of segments containing the set N in its interior.*

If we generalize the last definition to a case of countable coverings, we may have the set N of singularities of the function f which is not necessarily closed, and then it would be natural to consider countable coverings $\{\delta_i\}$ such that the function be summable on the complementary set. We thus arrive at the following definition investigated by Lusin [4], which generalizes Hahn's definition (see Section 6.4 and compare with Borel's definition).

The Lusin–Hahn–Lebesgue Definition

A function f is called integrable if (a) *there exists a set of singularities N of Lebesgue measure zero that is contained in open neighborhoods G of arbitrarily small measure, outside of which f is summable, and if* (b) *there exists the limit of the integral* (L) $\int_{CG} f(x)\ dx$ *as* $mG \to 0$.

*Assertion 2

A restricted Denjoy integral is more general than a Dirichlet–Lebesgue integral.

Proof

Let f be integrable in the Dirichlet–Lebesgue sense. Then P_1 (the set of nonsummability of f) is a countable set; totalization is completely accomplished by the successive application of Definitions 8.2 and 8.2′. The operation of totalization of a function integrable in the Dirichlet–Lebesgue sense is identical to the operation of constructing the Dirichlet–Cauchy integral described in Chapter 2 of this book (see the remark on p. 138).

An exact finite derivative with a set of points of nonsummability of positive measure is an example of a totalizable function which is non-integrable in the Dirichlet–Lebesgue sense.

Assertion 3

A restricted Denjoy integral is more general than the Lusin–Hahn–Lebesgue integral.

Proof

Formula (6.1), which is due to Hahn,[12] shows that a function integrable in the Lusin–Hahn–Lebesgue sense is totalizable. An example of a totalizable function not integrable in the Lusin–Hahn–Lebesgue sense can be obtained from the example of a function that is not Harnack-integrable but is Dirichlet-integrable (see Section 2.8) if in place of φ, in that example, a function is chosen that is conditionally Riemann-integrable on $[0, 1]$.*

---

[12] This formula is also valid in the case of the Lusin–Hahn–Lebesgue integral; here the integrals $\int_{a_\nu}^{b_\nu} f \, dx$ are interpreted in the Lebesgue's sense.

9 PERRON'S INTEGRAL

A new definition of the integral was proposed by Perron [1] in 1914. About ten years later, Alexandrov [1] and Looman [1] proved the equivalence of Perron's definition and the restricted Denjoy integral. However, Perron's definition is based on an idea which is completely different from Denjoy's initial idea and is connected with the concept of the integral as an increment of the primitive.

9.1 MAJOR AND MINOR FUNCTIONS

Before defining Perron's integral we must discuss several notions due to de la Vallée-Poussin [2]–[4]. These are the concepts of *major and minor* functions. The functions φ and ψ defined in Theorem 9.1 are major and minor functions, respectively.

Theorem 9.1 (de la Vallée-Poussin [2], [3])

For any function f summable on $[a, b]$ and $\varepsilon > 0$, there exist absolutely continuous functions φ and ψ, which possess the following properties:

160

(a) $\varphi(a) = \psi(a) = 0;$

(b) $\underline{D}\varphi \geqslant f, \overline{D}\psi \leqslant f$ *at all points at which* f *is finite;*

(c) $\varphi(x) \geqslant \int_a^x f(t) \, dt \geqslant \psi(x)$[1];

(d) $\varphi(x) - \int_a^x f(t) \, dt < \varepsilon, \quad \int_a^x f(t) \, dt - \psi(x) < \varepsilon.$

Proof

We first note that it is sufficient to prove the existence of a major function only, since the negative of a minor function of the function f is a major function of the function $-f$. Moreover, it is sufficient to consider the case of a nonnegative function f; indeed, consider the function f_N which is equal to f if $f(x) > -N$, and equal to $-N$ at other points where N is chosen so that

$$\int_a^b f_N(x) \, dx - \int_a^b f(x) \, dx < \frac{\varepsilon}{2}.$$

If φ_N is a major function of the nonnegative function $f_N(x) + N$, satisfying the conditions of Theorem 9.1 with ε replaced by $\varepsilon/2$, then $\varphi_N(x) - N(x - a) = \tilde{\varphi}(x)$ is a major function of the function f_N and, hence, *a fortiori* a major function of function f; moreover,

$$\tilde{\varphi}(x) > \int_a^x f(x) \, dx$$

and

$$\tilde{\varphi}(x) - \int_a^x f(x) \, dx = \varphi_N(x) - \int_a^x (f_N + N) \, dx$$

$$+ \int_a^x f_N(x) \, dx - \int_a^x f(x) \, dx < \varepsilon.$$

Thus, let f be a nonnegative summable function and let $e_l \stackrel{\text{def}}{=} E_x(l\eta \leqslant f(x) < (l+1)\eta)$, where η is an arbitrary positive number and $l = 0, 1, 2, 3, \ldots.$

We then have

$$\sum_0^\infty l\eta \, m e_l \leqslant (\text{L}) \int_a^b f(x) \, dx \leqslant \sum_0^\infty (l+1)\eta \, m e_l.$$

---

[1] Property (c) follows from (a) and (b) in view of the absolute continuity of φ and ψ and the theorem on differentiability of indefinite Lebesgue integrals.

Let $\{\varepsilon_i\}$ be a sequence of positive numbers and $\sum_0^\infty (l+1) \cdot \eta \varepsilon_l < \varepsilon/2$. Define for each e_l an open set G_l, such that $G_l \supset e$ and $mG_l - me_l < \varepsilon_l$. Let $e_l(x) \overset{\text{def}}{=} e_l \cdot [a, x]$, $G_l(x) \overset{\text{def}}{=} G_l \cdot [a, x]$. Then a fortiori $mG_l(x) - me_l(x) < \varepsilon_l$ and each of the functions me_l, mG_l is a monotonic absolutely continuous function on $[a, b]$. Put

$$\varphi(x) \overset{\text{def}}{=} \sum_{l=0}^\infty (l+1)\eta mG_l(x). \tag{9.1}$$

Evidently, $\varphi(x) \geqslant \int_a^x f(x)\, dx$. Since the general term of the series (9.1) differs from the general term of the series $\sum_0^\infty (l+1)\eta ml(x)$ by less than $l\eta \varepsilon_l$, it follows that the series (9.1) is uniformly convergent and

$$\varphi(x) - \int_a^x f(x)\, dx < \varepsilon$$

for η sufficiently small; moreover, φ is absolutely continuous. Indeed, a finite sum of the terms of series (9.1) is absolutely continuous; the increment of the remainder of the series (9.1) on any system of segments does not exceed (in view of monotonicity) its increment on the whole segment $[a, b]$, which in turn tends to zero.

It remains to show that $\underline{D}\varphi \geqslant f$ everywhere. Let $x \in e_l$, then $x \in G_l$ and if $h > 0$, $k > 0$ are sufficiently small, then $(x - k, x + h) \subset G_l$. We have

$$r(\varphi, x - k, x + h) \geqslant r((l+1) \cdot \eta mG_l(x), \quad x - k, \quad x + h).$$

(The right-hand side of this inequality is obtained from the left-hand side by omitting some positive terms.) The right-hand side of the inequality is equal to

$$(l+1)\eta \frac{k+h}{k+h} = (l+1)\eta$$

hence,

$$\underline{\lim}\, r(\varphi, x - k, x + h) \geqslant (l+1)\eta \geqslant f(x). \qquad \text{Q.E.D.}$$

Thus the function satisfies all the conditions of theorem.[2]

---

[2] De la Vallée-Poussin has shown how the method of major and minor functions allows us to obtain an elegant proof of the differentiability almost everywhere of absolutely continuous functions and functions of bounded variation without using Vitali's covering theorem.

De la Vallée-Poussin [2], [3] defined only *absolutely continuous* major and minor functions for a given function f and only for the case of a summable f. He also required that the inequalities

$$\underline{D}\varphi \geqslant f, \qquad \bar{D}\psi \leqslant f \qquad (9.2)$$

be valid at all points at which f is finite, i.e., almost everywhere. Clearly, if f is bounded, then the inequality (9.2) is satisfied *everywhere*.

Finally, de la Vallée-Poussin [4] defines a major (minor) function of an integral $\int_a^x f(t)\, dt$ to be an absolutely continuous set function $\Phi(E)$ ($\Psi(E)$) such that $\underline{D}\Phi > f$ ($\bar{D}\Psi < f$) at all points at which $f < +\infty$ ($f > -\infty$). It follows from de la Vallée-Poussin's argument that $\underline{D}\Phi \geqslant f$ ($\bar{D}\Psi \leqslant f$).

9.2   PERRON'S INTEGRAL

The notions of major and minor functions in de la Vallée-Poussin's investigations appeared only as details associated with absolutely continuous functions. Perron's integral is based on the elevation of these details to the level of a fundamental point.

Perron [1] introduces the notion of major and minor functions (*Oberfunktion, Unterfunktion*) for an *arbitrary* bounded function.

Definition 9.1

Let f be a bounded function on $[a, b]$. A continuous function φ (ψ) is called a major (minor) function of the function f if $\varphi(a) = \psi(a) = 0$ and if relation (9.2) is satisfied everywhere on $[a, b]$.

For the succeeding text, the following assertions concerning functions φ and ψ will be required.[3]

Assertion 1

The difference $\varphi - \psi$ is monotonically nondecreasing.

---

[3] In our opinion, it is unnecessary to indicate the priority of various authors concerning Assertions 1–10. These assertions could well have been rediscovered by various authors independently as the need arose.

Proof

We use the fact that if $\bar{D}f(x) \geqslant 0$ everywhere on $[a, b]$, then f is monotonically nondecreasing (a proof of this assertion can be obtained by the method of chains of intervals, see Section 4.7). The proof of Assertion 1 follows from the inequality $\bar{D}(\varphi - \psi) \geqslant \underline{D}\varphi - \bar{D}\psi \geqslant 0$.

Assertion 2

Let φ_1, φ_2, (ψ_1, ψ_2) be a major (minor) functions. Then $\varphi = \min(\varphi_1, \varphi_2)$ $[\psi = \max(\psi_1, \psi_2)]$ is a major (minor) function.

The proof follows directly from Definition 9.1.

Assertion 3

If f is continuous, then $\bar{D}f$, $\underline{D}f$ are L-measurable functions.

(A proof is given, for example, by Saks [1].)

Definition 9.2   (Perron [1])

A bounded function f on $[a, b]$ is called integrable in the Perron sense (P-integrable) if $\inf \varphi(b) = \sup \psi(b)$, where the infimum (supremum) is taken over all major (minor) functions φ (ψ). The P-integral is defined by the equality

$$(P) \int_a^b f(x)\, dx \stackrel{\text{def}}{=} \inf \varphi(b) = \sup \psi(b). \tag{9.3}$$

Perron did not aim at generalizing Lebesgue's integral; apparently at that time he was not aware of the generality of his integral. In his words, Perron introduced a definition of an integral (of a bounded function) whose basic properties were very simple to establish. We now list these properties in the form of Assertions 4–10 (Perron [1]). (Proofs of Assertions 4–10 are given by Natanson [1].)

Assertion 4

If f is a P-integrable function on $[a, b]$, then it is P-integrable on $[\alpha, \beta]$ for any subinterval $[\alpha, \beta] \subset [a, b]$.

Assertion 5

$$\int_a^c = \int_a^b + \int_b^c.$$

Assertion 6

$$\int kf = k \int f.$$

Assertion 7

$$\int f_1 + f_2 = \int f_1 + \int f_2.$$

Assertion 8

$$\int f_1 \geqslant \int f_2 \qquad \text{if} \quad f_1 \geqslant f_2.$$

Assertion 9

If f' is the (bounded)[4] derivative of f, then f' is P-integrable and $\int_a^b f' = f(b) - f(a)$.

Assertion 10

The function

$$F(x) \stackrel{\text{def}}{=} (\text{P}) \int_a^x f \, dx$$

is continuous with respect to x and $F'(x) = f(x)$ if f is continuous at the point x.

In concluding his paper, Perron notes that in view of de la Vallée-Poussin's theorem, every bounded L-integrable function is P-integrable. Moreover, he points out that his definition can be carried over without modification to the case of unbounded functions f with the additional

---

[4] The condition of boundedness cannot be removed (see, e.g., Natanson [1], Chapter XVI). (*Translator's note.*)

requirement of the existence of major and minor functions. Moreover, Assertions 1–10 remain valid.

As we shall see below, this definition makes sense in the case when f is finite everywhere; in the more general case, however, where f is finite almost everywhere, the additional requirements $\underline{D}\varphi > -\infty$ and $\overline{D}\psi < +\infty$ are needed. In general these requirements cannot be relaxed.

*Perron's definition is a definition of an integral in terms of a primitive function (this is our terminology for those definitions in which the integral is defined as an increment of a primitive) in the most general sense of the word. We have already noted (at the end of Part I of this book) that in the case of integration of continuous functions, an indefinite integral coincides completely with an increment of a primitive, as expressed by the equality

$$\int_a^b f(x)\, dx = F(b) - F(a). \tag{9.4}$$

A discontinuous function does not generally possess a primitive in the restricted sense of the word, therefore, a suitable generalization of the right-hand side of (9.4) is required. The assumption that f is a derived number for F does not help much in this case: one should bear in mind that the derivative as well as (Dini's) derived numbers possess certain properties which may not be shared by a given a priori discontinuous function.

But if one cannot find a function whose derived number equals f, then functions F_1 and F_2 may possibly exist whose derived numbers satisfy the inequalities

$$f(x) \leqslant \underline{D}F_1, \qquad f(x) > \overline{D}F_2.$$

In such a case, if $\inf(F_1(b) - F_1(a))$ coincides with the $\sup(F_2(b) - F_2(a))$, it is natural to define this common value to be the integral $\int_a^b f\, dx$. Such functions F_1 and F_2 are actually major and minor functions.

Thus, Perron's definition is an extension of the definitions of integrals in terms of primitives. It is thus a generalization of the Cauchy integral, Dirichlet integral, and Duhamel–Serre integral (the latter is discussed by Lebesgue [3]–[5], involving the idea of approximations).*

9.3 REFINEMENTS

In 1915, Bauer [1] analyzed Perron's integral for functions of several variables in detail and showed that every bounded P-integrable function is Lebesgue-integrable. The converse was proved by de la Vallée-Poussin. We thus have Theorem 9.2.

Theorem 9.2

In the class of bounded functions L-integration and P-integration are equivalent.

The idea of the proof of Bauer's theorem is as follows: (a) using Assertion 2, monotonic sequences $\{\varphi_n\}$ ($\{\psi_n\}$) of major (minor) functions are constructed such that $\varphi_n(b)\downarrow \inf \varphi(b)$, $\psi_n(b)\uparrow \sup \psi(b)$. If f is assumed to be P-integrable, then $\lim \varphi_n(b) = \lim \psi_n(b)$. It follows from Assertion 1 that $\lim \varphi_n(x) = \lim \psi_n(x)$, $a \leqslant x \leqslant b$. Hence, the sequences $\{\varphi_n\}$, $\{\psi_n\}$ converge to the same limit, which is the continuous function F. (b) Since the differences $\varphi_n(x) - F(x)$, $F(x) - \psi_n(x)$ are also monotonically nondecreasing and tend uniformly to zero for $n \to \infty$, it follows that the derived numbers of the function F are bounded and, moreover, equal almost everywhere to f. Hence, f is measurable and F is its indefinite Lebesgue integral.

Bauer also formulated a meaningful definition of major and minor functions of an unbounded function f (which may take infinite values at certain points) by stipulating the additional requirement $\underline{D}\varphi > -\infty$, $\overline{D}\psi < +\infty$.[5]

Definition 9.3

Let f be an arbitrary function on $[a, b]$. Major and minor functions of the function f are the functions φ and ψ, respectively, which vanish at the point $x = a$ and which satisfy everywhere relations $\underline{D}\varphi \geqslant f$, $\overline{D}\psi \leqslant f$ and, moreover,

$$\underline{D}\varphi > -\infty, \qquad \overline{D}\psi < +\infty. \qquad (9.5)$$

---

[5] According to Bauer, these conditions were pointed out by W. Gross. They are necessary since otherwise ψ may turn out to be greater than φ.

Using these refined definitions of major and minor functions which are also applicable in the case of unbounded functions, Bauer formulated the general definition of Perron's integral.

Definition 9.4

A function f is called integrable in the Perron's sense if it possesses major and minor functions φ and ψ and if, moreover, inf $\varphi(b)$ = sup $\psi(b)$. The P-integral is defined by Eq. (9.3).

Corollary

An exact finite derivative is P-integrable.

Remark 1. The functions φ and ψ defined above satisfy the conditions stipulated in Assertions 1–3, and for the Perron integral of an arbitrary function, Assertions 4–10 are valid.

Remark 2. If $f(x_0) = +\infty$, then $\varphi'(x_0) = +\infty$; if $f(x_0) = -\infty$, then $\psi'(x_0) = -\infty$. Hence, in order that f possess a major (minor) function, it is necessary that $f(x) < +\infty$ $(f(x) > -\infty)$ almost everywhere. This assertion follows from a theorem proved by Lusin [1], [2], [4] which states that the derivative of a continuous function g can assume infinite values on at most a set of points of measure zero. It is possible that Bauer was not aware of Lusin's theorem. However, the particular case of this theorem when g is monotonic was known to him. And this fact was sufficient to assert that in order for f to possess major and minor functions, it is necessary that f be finite almost everywhere.

*The following questions naturally arise. Is there a measurable function, finite almost everywhere, which does not possess a major (minor) function? Is there a measurable function which possesses a major and minor function, but which is not P-integrable?

Consider a measurable finite positive nonsummable function f. If this function possesses a major function φ, then $\underline{D}\varphi \geqslant f > 0$ and φ is a monotonic function with a summable derivative which contradicts the assumption that f is nonsummable. (Thus, the answer to the first question is affirmative.)

The second question was resolved by J. Marcinkiewich in the negative (see Saks [1], p. 253).*

The greater generality of Perron's definition as compared with Lebesgue's definition follows from the theorem on the P-integrability of L-integrable functions, which is proved by Bauer [1]. This theorem is formally stronger than de la Vallée-Poussin's Theorem 9.1 since the conditions on major and minor functions in Definition 9.3 are more stringent than the conditions required by de la Vallée-Poussin ($\underline{D}\varphi \geqslant f$, $\overline{D}\psi \leqslant f$ everywhere). This theorem, however, is equivalent to de la Vallée-Poussin's theorem proved in [4] which was discussed on p. 163.

In concluding his paper, Bauer formulates a generalized definition of major and minor functions which differ from Definition 9.3 in that the inequalities $\underline{D}\varphi \geqslant f$, $\overline{D}\psi \leqslant f$ must be satisfied *almost everywhere*, and shows that the definition of P-integrability, on the basis of such an extended interpretation of major and minor functions, is not more general than Definition 9.4. Indeed, having, for example, a major function φ which satisfies the inequality $\underline{D}\varphi \geqslant f$ almost everywhere, we obtain a major function in the sense of Definition 9.3 by adding to φ a monotonic function φ_1 that posesses an infinite derivative on the set of point on which $\underline{D}\varphi \geqslant f$ is violated.

It became evident as a result of the works of de la Vallée-Poussin, Perron, and Bauer that Perron's integral is more general than Lebesgue's integral. This integral also reconstructs the primitive function and thus possesses the basic characteristic feature of Denjoy's integral.[6] The generality of Perron's integral was fully established a few years later when Hake, Alexandrov, and Looman proved the equivalence of integration in the restricted Denjoy sense with Perron's integration.

In 1921, Hake [1] showed that a function integrable in the restricted Denjoy sense is P-integrable. We now remark on certain specific features of this paper. In his definition of major and minor functions and P-integrability, Hake reproduces Bauer's definition with the only modification that in place of derivatives $\underline{D}\varphi$, $\overline{D}\psi$, right-hand derivatives $\underline{D}^+\varphi$, $\overline{D}^+\psi$ appear. (With this definition of the functions φ and ψ, Assertions 1–3 remain valid.)[7] Furthermore, Hake also formulates the definition of

[6] There is no mention of Denjoy's results in the papers quoted above.

[7] Hake remarks that he found out about Bauer's paper after his own paper was written.

P-integrability of a function using absolutely continuous major and minor functions and shows that such a definition is equivalent to the definition of L-integrability. (In one direction this assertion was proved as we have seen by de la Vallée-Poussin; the other direction follows immediately from the summability of f.)

The P-integrability of totalizable functions is proved in three stages: it is shown that a function integrable in the Dirichlet–Lebesgue sense is P-integrable; next it is shown that a function integrable in the Harnack–Lebesgue sense is P-integrable; and finally, the general theorem is proved, using transfinite induction, based on the fact that a restricted Denjoy integral is obtained as a result of a transfinite succession of integration operations in the Dirichlet–Lebesgue and Harnack–Lebesgue senses.

Hake concludes his paper with the remark that the totalizability of P-integrable functions is as yet unknown.

In 1924, Alexandrov [1] and, independently, Looman [1][8] in 1925 showed that a P-integrable function is totalizable in the strict sense. Thus the equivalence between Perron's integrals and restricted Denjoy's integrals was finally and completely established. (Proofs are given by Natanson [1] and Saks [1, p. 247].)

In connection with the definition of a P-integral and the Hake–Alexandrov–Looman theorem, a few words are in order concerning the problem of a constructive definition of Denjoy's integral without using transfinite numbers. The desirability of such a definition was suggested in particular by Lusin [4], [6]. The main objection to the utilization of transfinite induction in the construction of Denjoy's integrals is as follows: although for the totalization of each specific function an at most countable chain of sets P_γ (cf. Chapter 8) is required, γ may, however, be an arbitrarily large transfinite number of the second class; therefore it is necessary to have available for totalization of an *a priori* given totalizable function an uncountable set of transfinite numbers of the second class. In other words, Denjoy's process is substantially more complicated than that of Lebesgue. Besides, for the proof that the chain $\{P_\gamma\}$ terminates at a certain point (i.e., to prove the Cantor–Baire stationarity principle), it is also necessary to consider all transfinite numbers of the second class.

---

[8]Alexandrov's and Looman's papers were received by the editors of the corresponding journals almost simultaneously in June 1923.

It may seem that the definition proposed by Perron solves our problem. However, Looman as early as 1925 pointed out that Perron's definition is not constructive. (We have also noted above that, as a descriptive definition, Lusin's Definition 8.10 is preferable to Perron's definition.) The actual determination of major and minor functions is probably not simpler than the construction of a total. It also seems that the process of totalization is the simplest constructive process for the definition of Perron's integral as well. (See in this connection, N. K. Bari and D. E. Men'schov's comments on N. Lusin's [6] thesis. These commentaries are contained in the Lusin [6] reference.)

10 DANIELL'S INTEGRAL

In the second decade of this century, integration penetrated more and more into spaces differing from the initial prototype, the n-dimensional Euclidean space. This departure from the bounds of Euclidean spaces was dictated mainly by the development of functional analysis. It was awkward to associate integration of functions in general spaces with the properties of the elements and subsets of the space (among these properties the most important is the existence of a class of sets of a determined algebraic nature on which a measure is defined). In certain problems the interpretation of an integral as a set function became unsuitable and the view of an integral as a functional became preferable. This new tendency is expressed by Daniell's definition of the integral given in 1919.

10.1 DANIELL'S DEFINITION

Before presenting Daniell's ideas (Daniell [1]) we would like to point out that, while considering both concepts of the integral as a set function and as a functional, we do not intend to show the contrast between one and the other. On the contrary, we should emphasize that these two concepts supplement one another and are inseparably connected. This connection has already been expressed by Lebesgue in his axioms of integration. An integral is a set function as well as a functional and one can merely imply his preference for one of these two characteristic features of the integral and intentionally disguise the other under certain circumstances.

Lebesgue's point of view of an integral was strongly in favor of its interpretation as a set function. This attitude was justified historically, of course. However, it would have been sufficient to consider, in place of a measure, a nonnegative *functional* $U(f)$ defined on an additive class of linear combinations of characteristic functions of measurable sets on the interval $[0, 1]$, which possesses the homogeneity and additivity properties [i.e., $U(f) \geqslant 0$ if $f \geqslant 0$; $U(cf) = cU(f)$; $U(f_1 + f_2) = U(f_1) + U(f_2)$], normalized by the conditions $U(\chi_{E_1}) = U(\chi_{E_2})$ if E_1 and E_2 are congruent and $U(1) = 1$, and then to define

$$U(\varphi) \overset{\text{def}}{=} \lim_{n \to \infty} U(f_n) \qquad (10.1)$$

for every function φ which is the limit of a monotonically increasing sequence $\{f_n\}$ of step functions. A functional defined by Eq. (10.1) is precisely the Lebesgue integral of a nonnegative function. Here we have expressed a well-known fact using a different terminology: the starting point was not the notion of measure but of a functional defined on some initial sufficiently simple class of functions (linear combinations of characteristic functions).

If we were to generalize the notion of the integral by interpreting it as a functional in the same manner as Fréchet obtained a generalization of an integral by viewing it as a set function, it would be necessary to define *a priori* an additive class of functions and to define on it a nonnegative, homogeneous and additive functional $U(f)$, and then to repeat the well-known process of continuation of the functional by means of Eq. (10.1), for example. This was the method of generalization suggested by Daniell.

Daniell [1] considers an arbitrary space M; the initial class of functions T_0 is defined by the following conditions:

(1)   If $f_1, f_2 \in T_0$, then $f_1 + f_2 \in T_0$.

(2)   If $f \in T_0$, then $cf \in T_0$, then $cf \in T_0$, where c is a real number.

(3)   If $f_1, f_2 \in T_0$, then $\max(f_1, f_2) \in T_0$ and $\min(f_1, f_2) \in T_0$.

All the functions in the class T_0 are bounded.

The class of real-valued functions which admit a finite number of values with a finite number of discontinuity points is an example of a class satisfying the conditions above. It is also helpful to bear in mind that the subsequent construction of Daniell's integral will correspond fully to Young's construction for the Stieltjes integral (see Chapter 7) based on the method of monotonic sequences. For the purpose of the following discussion, we shall call the functions of the class T_0 "step" functions.

A functional $U(f)$ is then defined on the class T_0. The following conditions are imposed on $U(f)$:

(A) $U(f_1 + f_2) = U(f_1) + U(f_2)$.

(C) $U(cf) = cU(f)$, where c is a real number.

(L)   (Lebesgue's property)   If $f_1 \geqslant f_2 \geqslant f_3 \geqslant \cdots$ and $\lim_{n \to \infty} f_n = 0$, then $\lim U(f_n) = 0$.

(P) $U(f) \geqslant 0$ if $f \geqslant 0$.

Thus the "integral" $U(f)$ is defined for step functions. Conditions (A), (C), and (P) express the basic properties of a Riemann–Stieltjes integral $\int f \, d\varphi$ with respect to monotonically increasing function φ. Condition (L) is prepared "in advance" to assure uniqueness of the continuation of $U(f)$.

The next stage in the construction is the continuation of $U(f)$ by Eq. (10.1) on the class T_1 of functions which are limits of monotonically increasing sequences of functions of the class T_0. The independence of this continuation of the sequence of functions is guaranteed by the condition (L). The class T_1 obviously contains the class T_0. [In Young's construction, the procedure of continuation of $U(f)$ to the functions of the class T_1 corresponds to the continuation of the integral from the class of step functions to the class of functions semicontinuous from below (the class of l-functions).][1]

---

[1] We note that functions f, which are limits of monotonically *decreasing* sequences of functions in the class T_0, can be defined by the requirement that $-f \in T_1$.

Finally, the last stage in the construction of an integral of an arbitrary function is the definition of the upper and lower integrals

$$\dot{U}(f) \overset{\text{def}}{=} \inf_{\substack{\varphi \in T_1 \\ \varphi \geqslant f}} U(\varphi), \qquad \underline{U}(f) \overset{\text{def}}{=} -\dot{U}(-f). \tag{10.2}$$

A function is called summable if $\dot{U}(f) = \underline{U}(f)$, and if, moreover, $\dot{U}(f)$ and $\underline{U}(f)$ are finite. In this case, $U(f) \overset{\text{def}}{=} \dot{U}(f)$; a function f is summable if and only if $|f|$ is summable. The class of summable functions possesses properties (1)–(3), which are valid for the class T_0, and moreover, the functional $U(f)$ on the class of summable functions possesses properties (A), (C), (L), and (P). Furthermore, the following "Lebesgue theorems" are valid: (a) If $\{f_n\}$ is a monotonic sequence of summable functions and if $\lim_{n\to\infty} U(f_n) \neq \pm\infty$, then $\lim_{n\to\infty} f_n$ is summable and $\lim_{n\to\infty} U(f_n) = U(\lim_{n\to\infty} f_n)$; (b) if $\{f_n\}$ is a convergent sequence of summable functions and if $|f_n| \leqslant \varphi$ where φ is summable, then $\lim_{n\to\infty} f_n$ is summable and $\lim_{n\to\infty} U(f_n) = U(\lim_{n\to\infty} f_n)$.

10.2 THE GENERAL CASE

The model discussed above corresponds to the case when $U(f)$ is a positive functional, for example, Lebesgue's integral or Stieltjes' integral with a positive generating function. What modifications are required in this model if $U(f)$ does not satisfy condition (P) on T_0 [which is the case, for example, when $U(f) = \int f \, d\varphi$ and φ is not monotonic]? In this event we follow the usual method of decomposing the functional $U(f)$ into its positive and negative parts and each one of these parts is a positive functional. The decomposition is usually carried out by decomposing the measure (i.e., the function φ in Stieltjes' integral $\int f \, d\varphi$) into two nonnegative measures. However, this method is not directly applicable in our case since the functional $U(f)$ does not depend explicitly on a measure. Some additional modifications are thus required.

Daniell therefore considers a functional $U(f)$ which satisfies all the conditions stated above except for condition (P). The latter is replaced by the following:

(M)   There exists a finite functional $M(f)$ defined for all positive functions, satisfying the condition $M(\varphi) \leqslant M(f)$ provided $\varphi \leqslant f$, and such that $U(f) \leqslant M(|f|)$ for any $f \in T_0$.[2]

The functional satisfying conditions (A), (C), (P), and (L) will also satisfy condition (M) with $M(f) = U(|f|)$. The decomposition into positive and negative parts of $U(f)$ is as follows: if $f \geqslant 0$, then

$$U^+(f) \overset{\text{def}}{=} \sup_{0 \leqslant \varphi \leqslant f} U(\varphi).$$

In the general case when $f = f^+ - f^-$, then[3]

$$U^+(f) \overset{\text{def}}{=} U^+(f^+) - U^+(f^-)$$

and

$$U^-(f) \overset{\text{def}}{=} U^+(f) - U(f).$$

$U^+(f)$ and $U^-(f)$ satisfy conditions (A), (C), (P), and (L). Using the method described above, the continuation of the functionals $U^+(f)$, $U^-(f)$ to the classes of summable functions (relative to these functionals) is constructed. The intersection of these classes is nonempty, since each one contains the class T_1. The function f is called summable if it is summable relative to $U^+(f)$ and $U^-(f)$. In this case

$$U(f) \overset{\text{def}}{=} U^+(f) - U^-(f).$$

We have thus constructed the required generalization, which (as was pointed out by Daniell) includes the Lebesgue and Stieltjes integrations —these correspond to particular choices of the class T_0 and of the functional $U(f)$ defined on this class.

We emphasize once again that Daniell's method reproduces exactly Young's method from which certain inessential details have been excluded; moreover, in only a few cases new methods of proofs are required.[4]

[2] If $U(f) = \int f\, d\varphi$, M may be defined by $M(\psi) = \int |\psi|\, d\varphi$.
[3] f^+ and f^- also belong to T_0 since $f^+ = \max(f, 0), f^- = -\min(f, 0)$.
[4] It is noteworthy to point out that Young eliminated the notion of measure from his definitions in an effort to make it more similar to the previously known methods and to make it more applicable to the "mathematicians of the older school." Daniell also excludes the notion of measure but his reasons in this connection are completely opposite.

In concluding his paper, Daniell [1] points out that if one starts the construction of an integral from a *measure* defined on a class of sufficiently simple sets, then linear combinations of characteristic functions should be chosen for T_0 and $U(f)$ should be defined as a linear combination of the corresponding measures. In [2], Daniell constructs specific examples of integrals in infinitely dimensional spaces (Fréchet spaces).

It follows from the above that the procedure for constructing Daniell's integral is a reformulation of well-known constructions utilizing the notion of measure. Is it possible conversely to interpret Daniell's integral as an integral with respect to some measure or as a measure in general? In any case the following interpretation arises naturally: transform the functional $U(f)$ into a set function on the space of pairs (y, p) where y is a real number and p is an element of the initial space M by considering $U(f)$ to be a set function defined on the ordinate sets $E_{(y, p)}$ ($0 \leqslant y < f, p \in M$) of positive functions in the class T_0. [The class of ordinate sets, on which the measure $U(f)$ is initially defined, thus possesses a specific algebraic structure.] Then, one can assert that the continuation of the functional $U(f)$ to the class of summable functions corresponds essentially to the choice of sets measurable (ordinate) with respect to the outer measure constructed by the set function $U(f)$.

In Daniell's other paper [3] the notions of measure and of a measurable function are introduced on the space M in terms of the functional $U(f)$. On the basis of these notions, a closer analogy between Daniell's and Lebesgue's theories can be established.

In subsequent years, Daniell's integral underwent various modifications and a number of different versions of its construction are available at present. (The theory of Daniell's integral is treated in detail in A. C. Zaanen's book [1]; this book also contains a comprehensive bibliography.)

CONCLUSION

We hope that our objective of tracing the development of the two basic interpretations of the integral, as an increment of a primitive and as a limit of sums, has been achieved. Now we are in a position to comment with greater justification on the individual features of each of these approaches.

As long as these two concepts were applied to continuous functions, they were equivalent; the equivalence ceased to exist in the case of discontinuous functions. However, no matter how the content of these concepts changed, the essential characteristics initially intrinsic in each of them were preserved. It was thus required that the primitive should (a) possess properties linking it very closely with the integrand, to the extent that these properties would assure its uniqueness; and (b) coincide with the classical derivative in the largest possible number of cases. Definitions coresponding to the first concept were necessarily descriptive and the corresponding integrals were, as a rule, only conditionally convergent.

Moreover, no matter what changes the second concept underwent, it was required that (a) the process of formation of integral sums be

178

regular and independent of the function to which it was applied, and (b) in all cases in which the ordinate set is defined, the integral be interpreted as the area of the ordinate set. Here the corresponding definitions are constructive and the integrals, as a rule, are absolutely convergent.

In all cases, however, it is desirable that the integral possess certain basic properties as a functional and as a set function.

The first concept reached its peak with the Denjoy integral in its descriptive definition. The Fréchet integral is essentially the completion of the second concept. (It is supplemented by the investigations of Riesz, Daniell, and Banach, which were carried out from the functional analysis point of view.)

The obstacles preventing investigators in the 19th century from predicting the future development of these concepts disappeared in the wake of the "explosion" (this, in our opinion, is the proper term to describe the new discoveries and trends whose main leaders were Borel, Lebesgue, and Young) that took place at the beginning of the 20th century. As time elapsed the power of this explosion was universally acknowledged.

In the 1920's the confines of both concepts were determined—the techniques of the metric theory of functions became an active tool for investigators in this area. (Although, some papers of a methodological nature continued to appear aimed at simplifying the theory of Lebesgue's integral mainly by eliminating the notion of measure. This research overshadowed the fundamental characteristic of the Lebesgue integral, namely, that it is the sole solution to the integration problem. We have already mentioned that no other definitions can eliminate the basic difficulty involved in the noncountability of the process.)

As nothing noteworthy can be expected in the future for both concepts of integration, this would have been an appropriate point at which to conclude our exposition. Nevertheless, we shall bow to tradition by making a number of necessarily sketchy concluding remarks, concerning the further development of integration theory (in the 1920's and 1930's).

We begin with the first concept. Lusin [1], [4], [6] extended the notion of the primitive F of a function f by interpreting it as a continuous function differentiable almost everywhere and possessing f as its derivative almost everywhere. He also proved the fundamental theorem on the existence of a primitive for every measurable function, finite almost everywhere, and searched for additional conditions to assure the uniqueness of the primitive and, at the same time, to make it more general as compared with the (restricted) Denjoy integral.

In the course of this investigation the N-property of functions and a more stringent property termed by Lusin the zero variation was studied. However, the results were not completely successful and it was discovered later that primitives possessing these properties are not unique. (Details are given in Lusin [6] (1951), which is an extension and commentary on his thesis [4].)

The most important external stimulant in the development of integration was the desire to express the coefficients of a trigonometric series in terms of its sum using Fourier's formulas—the so-called Fourier problem. Here we witness how the effect becomes the cause at a certain stage of development: Lusin observed that the expansion of a given function in a trigonometric series (converging to the function in a certain sense) may serve as an origin of the definition of its primitive. Let, for example, the series $\sum a_n \cos nx + b_n \sin nx$ converge almost everywhere to f. Formally integrating this series, we obtain $\sum (-b_n/n) \cos nx + (a_n/n) \sin nx$, which is convergent almost everywhere. It is then natural to attempt to define its sum as the primitive of the function f.

However, as further investigations by Lusin have shown (see [6], commentaries), the obstacle is that there exists a series with nonzero coefficients which converges almost everywhere to zero, while the series obtained by term-by-term integration fails to possess this property (D. E. Men'shov's example). "Trigonometric" totalizations are discussed by S. Verblunksy, *Fund. Math.* **23** (1934); A. Zygmund, *Fund. Math.* **26** (1936); J. Ridder, *Math. Z.* **42**, 234–269 (1937).

Thus we are faced with one direction in the development of the concept of totalization based on its interpretation as an operation allowing us to determine the coefficients of a trigonometric series by its sum.

If we take into account the fact that totalizations were conceived with the aim of determining the primitives, it will be clear that other stimulants in their development were the generalizations of the notion of the derivative (and not only the first derivative). In 1921 Denjoy constructed a totalization which determined a function by means of its second Schwarz derivative and conjectured the possibility of determining a primitive from its symmetric derivative. [*Compt. Rend.* **172** (1921), **173** (1921); see also *Fund. Math.* **25** (1935)]. He described the corresponding totalization in 1956 [*Compt. Rend.* **241** (1955)]. Perron's definition was also developed by several authors. First, major and minor functions, defining an integration equivalent to the general totalization in the Denjoy–Khinchin sense, were constructed [J. Ridder, *Math. Z.* **34** (1931–1932); M. D. Kennedy and S. Pollard, *Math. Z.* **39**, (1935)]. The fact that these major and minor functions are determined by means of the approximate derivative numbers is not surprising. Classes of major and minor functions leading to the Riemann and Lebesgue integrals were also constructed.

Functions that are approximately differentiable at every point are, in general, only approximately continuous. This resulted in a study of totalizations leading to approximately continuous functions [J. C. Burkill, *Math. Z.* **34**, 270–278, (1931–1932); J. Ridder, *Fund. Math.* **22**, 163–179 (1934). We note that it was Lusin [4] and [6] and Young [7] who earlier suggested the advisability of investigating discontinuous primitive functions.

In Leçons II, Lebesgue pointed out the modifications to be included in the general procedure in order to obtain the Denjoy–Stieltjes integral.

We now discuss briefly various developments of the second concept of the integral.

In 1907, Hellinger (in his dissertation at Göttingen) investigated quadratic forms in a countable number of variables and guided by one general idea, arrived at several definitions of the integral. One of his typical definitions is now presented.

Hellinger's Definition

Let f, g, h be continuous functions on $[a, b]$ and let g and h be strictly monotonically increasing. Under the condition that $[f(\Delta)]^2 \leqslant g(\Delta)h(\Delta)$ for every $\Delta \subset [a, b]$, there exists the limit

$$\lim_{d(\sigma) \to 0} \sum \frac{[f(\Delta_i)]^2}{g(\Delta_i)},$$

called the integral $\int_a^b (df)^2/dg$.

This integral is additive when considered as a function of a segment and is continuous as a function of its upper limit. In particular, the variation of a continuous function is a Hellinger integral. Clearly, one can also study integrals of the type $\int \varphi((df)^2/dg)$; the case when $f(\Delta)$ and $g(\Delta)$ are set functions was investigated by Radon [1]; the general procedure for constructing such an integral was pointed out by Lebesgue in Leçons II.

Here we are dealing, probably for the first time, with the following situation. A function of a segment is considered, say $l(\Delta)$, which is generally nonadditive; the function is partitioned in a certain domain and the existence of the limit of sums arising from these partitions is required (in Hellinger's case, $l(\Delta) = [f(\Delta)]^2/g(\Delta)$. This problem was considered by Burkill, who initially defined an integral as the $\lim \sum_i l(\Delta_i)$ [J. C. Burkill, *Proc. London Math. Soc.* **22**, 275–310 (1924).] The next step is to interpret the elements of the integral sums $f(\xi_i) \cdot mE_i$ as the values of a certain multivalued set function $\chi(E) = l \cdot mE$, where l is an aribtrary value intermediate between the supremum and the infimum of the function to be integrated. If we now disregard the function of a point, then a multivalued set function $\chi(E)$ will remain and the integral becomes the limit of the sums $\sum_i \chi(E_i)$. Finally, the last step is to consider the integral sums as functions of a partition, and their limit is interpreted correspondingly (such a limit was defined by Pollard, see Chapter 4). The notion of the limit of a function of partitions was generalized thereafter by E. H. Moore and H. J. Smith [*Amer. J. Math.* **44** (1922)] and by S. O. Shatunovskiĭ ["An Introduction to Analysis," Odessa, 1924].

Thus, let sequences of sets (called partitions) be defined in a space of an arbitrary nature and let a set function Φ be defined on the elements of these partitions.

The integral $\int_E d\Phi$ is defined as the limit of the function of a partition $\sum \Phi(E_i)$ constructed on the partitions of the set E. This very general scheme for constructing an integral (described here without details) is due to A. N. Kolmogorov [*Math. Ann.* **103** (1930)]. It encompasses the Lebesgue–Stieltjes integral, Fréchet integral, and so on.

A fairly general theory was also constructed by Ridder. He utilized the idea of defining an integral as a measure of an ordinate set in a certain product space [J. Ridder, *Fund. Math.* **24** (1935)].

Finally, with the development of the theory of linear spaces it became necessary to define integrals of functions with values in general linear spaces. Clearly, in this case the integral sums are linear combinations of elements in this space. Therefore, these sums, as well as the integral, are elements in the same space. Such an integral was constructed by S. Bochner [*Fund. Math.* **20**, 262–276 (1933)]. More details are given in the book by E. Hille and R. Phillips: "Functional Analysis and Semigroups," Vol. 31, Colloquium Publ., Amer. Math. Soc., Providence, Rhode Island, 1957.

In conclusion, it should be observed that certain definitions of an integral arising in connection with specific problems have not been mentioned at all in this brief survey. (Integrals that have applications in the theory of trigonometric series are discussed in N. K. Bari's book: "Trigonometric Series," Fizmatgiz, Moscow, 1961 (English transl. in 2 vols. entitled: "A Treatise on Trigonometric Series," Macmillan, New York, 1964).

REFERENCES

ALEXANDROV, P. S.
    [1] Über die Äquivalenz des Perronschen und des Denjoyschen Integralbegriffes. *Math. Z.* **20**, 213–222 (1924).

BANACH, S.
    [1] Sur le problème de la mesure. *Fund. Math.* **4**, 7–33 (1923).

BAUER, H.
    [1] Der Perronsche Integralbegriff und seine Beziehung zum Lebesgueschen. *Monatsh. Math. Phys.* **26**, 153–198 (1915).

BOKS, T. J.
    [1] Sur les rapports entre les méthodes d'intégration de Riemann et de Lebesgue. *Rend. Circ. Mat. Palermo* **45**, 211–262 (1921).

BOREL, E.
    [1] "Leçons sur la théorie des fonctions." Paris, 1898.
    [2] "Leçons sur la théorie des Fonctions." 2nd ed. Paris, 1914.
    [3] Un théorème sur les ensembles mesurables. *Compt. Rend.* **137**, 966–969 (1903).
    [4] "Leçons sur les fonctions de variables réelles." Paris, 1905.
    [5] Sur la définition de l'intégrale définie. *Compt. Rend.* **150** 375–378 (1910).

[6] Sur une condition générale de l'intégrabilité. *Compt. Rend.* **150**, 508–510 (1910).

[7] Le calcul des intégrales définies. *Math.* **8**, (6), 159–210 (1912).

(8) Sur l'intégration des fonctions non bornées et sur les définitions constructives. *Ann. Ecole Norm. Sup.* **36**, 71–91 (1919)

[9] A propos de la définition de l'intégrale définie. *Ann. Ecole Norm. Sup.* **37**, 461–462 (1920).

[10] Sur les théorèmes fondamentaux de la théorie des fonctions de variables réelles. *Compt. Rend.* **154**, 413–415 (1912).

CANTOR, G.

[1] Über unendliche lineare Punktmannichfaltigkeiten, 4. *Math. Ann.* **21**, 51–58 (1883).

[2] Über unendliche lineare Punktmannichfaltigkeiten, 6. *Math. Ann.* **23**, 453–488 (1884).

[3] De la puissance des ensembles parfaits de points. *Acta Math.* **4**, 381–392 (1884).

CARATHÉODORY, C.

[1] Über das lineare Mass von Punktmengen-eine Verallgemeinerung des Längebegriffs. *Nachr. Ges. Wiss. Göttingen* 404–426 (1914).

CAUCHY, O.

[1] "Résumé des Leçons Données à l'Ecole Royale Polytechnique, le Calcul Infinitésimal," Vol. 1. Paris, 1823.

[2] Russian translation of [1]. St. Petersburg, 1832.

DANIELL, P. J.

[1] A general form of integral. *Ann. Math.* **19**, 279–294 (1917–1918).

[2] Integrals in an infinite number of dimensions. *Ann. Math.* **20**, 281–288 (1918).

[3] Further properties of the general integral. *Ann. Math.* **22**, 203–220 (1920).

DARBOUX, G.

[1] Mémoire sur les fonctions discontinues. *Ann. Ecole. Norm. Sup.* **4**, (2) 57–112 (1875).

DE LA VALLÉE-POUSSIN, C. J.

[1] Recherches sur la convergence des intégrales définies. *J. Math.* **8** (4), 421–467 (1892).

[2] "Cours d'Analyse Infinitésimale," Vol. 1, 3rd ed. Louvain, Paris, 1914.

[3] Russian translation of [2]. Odessa, 1922.

[4] "Intégrales de Lebesgue, Fonctions d'Ensemble. Classes de Baire." Gauthier-Villars, Paris, 1916.

DENJOY, A.

[1] Une extension de l'intégrale de M. Lebesgue. *Compt. Rend.* **154**, 859–862 (1912).

[2] Calcul de la primitive de la fonction dérivée la plus générale. *Compt. Rend.* **154**, 1075–1078 (1912).

[3] Sur la dérivation et son calcul inverse. *Compt. Rend.* **162**, 377–380 (1916).
[4] Mémoire sur la totalisation des nombres dérivés non sommables. *Ann. Ecole. Norm. Sup.* **33** (3), 127–222 (1916).
[5] Totalisation des nombres dérivés non sommables. *Ann. Sci. Ecole Norm. Sup.* **34** (3), 181–236 (1917).
[6] Sur l'intégration riemaniènne. *Compt. Rend.* **169**, 219–221 (1919).
[7] Sur la définition riemaniènne de l'intégrale de Lebesgue. *Compt. Rend.* **193**, 695–698 (1931).

DU BOIS-REYMOND, P.
[1] Versuch einer Classification der willkürlichen Funktionen reeller Argumente nach ihren Aendrungen in den kleinsten Intervallen. *J. Math.* **79**, 21–37 (1875).
[2] Der Beweis des Fundamentalsatzes der Integralrechnung. *Math. Ann.* **16**, 115–128 (1880).
[3] " Die allgemeine Funktionentheorie I." Tübingen, 1882.

ENZYKLOPÄDIA DER MATHEMATISCHEN WISSENSCHAFTEN, Vol. II, Part III, *Leipzig*, 1923–1928. Rosenthal's article, "Neuere Untersuchungen über Funktionen reeller varänderlichen."

FRÉCHET, M.
[1] Sur l'intégrale d'une fonctionelle étendue à un ensemble abstrait. *Bull. Soc. Math. France* **43**, 249–267 (1915).
[2] On Pierpont's definition of integrals. *Bull. Amer. Math. Soc.* **22**, No. 6, 295–298 (1916).
[3] La vie et oeuvre d'Emile Borel. *Ensèignement* **11**, (2) 1–94 (1965).

FREUD, P.
[1] Über die uneigentlichen bestimmten Integrale. *Monats. Math. Phys.* **16**, 11–24 (1905).

HAHN, H.
[1] Über eine Verallgemeinerung der Riemannschen Integraldefinition. *Monats. Math. Phys.* **26**, 3–18 (1915).
[2] Über Annäherung der Lebesgueschen Integrale durch Riemannsche Summen. *Sitzber. Akad. Wiss. Wien., Abt. IIa*, **123**, 713–743 (1914).

HAKE, H.
[1] Über de la Vallée-Poussins Ober- und Unterfunktionen einfacher Integrale und die Integraldefinition von Perron. *Math. Ann.* **83**, 119–142 (1921).

HANKEL, H.
[1] Untersuchungen über die unendlich oft oscillierenden und unstetigen Funktionen. *Math. Ann.* **20**, 63–112 (1882).

HARNACK, A.
[1] Vareinfachung der Beweise der Theorie der Fourierschen Reihe. *Math. Ann.* **19**, 235–279, (1882).
[2] " Differenzial- und Integralrechnung." Leipzig, 1881.

[3] Über den Inhalt von Punktmengen. *Math. Ann.* **25**, 241–250 (1885).

[4] Anwendung de Fourierschen Reihe auf die Theorie der Funktionen einer komplexen Veränderlichen. *Math. Ann.* **21**, 305–327 (1883).

[5] Die allgemeinen Sätze über den Zusammenhang der Funktionen einer reellen Variabel mit ihren Ableitungen II. *Math. Ann.* **24**, 217–252 (1884).

HAUSDORFF, F.
[1] "Grundzüge der Mengenlehre." Leipzig, 1914.

HILDEBRANDT, T. H.
[1] On integrals related to and extensions of the Lebesgue Integrals. *Bull. Amer. Math. Soc.* **24**, 113–144, 177–202, (1917–1918).

HÖLDER, O.
[1] Zur Theorie der trigonometrischen Reihen. *Math. Ann.* **24**, 181–216 (1884).

JORDAN, C.
[1] Remarques sur les intégrales définies. *J. Math.* **8**, (4), 69–99, (1892).
[2] "Cours d'Analyse," Vols. I, II, 2nd ed. Paris, 1893–1894.

KEMPISTY, S.
[1] Un nouveau procédé d'intégration des fonctions mesurables non sommables. *Compt. Rend.* **180**, 812–815 (1925).
[2] Sur l'intégrale (A) de M. Denjoy. *Compt. Rend.* **185**, 749–751 (1927).

KHINCHIN (KHINTCHINE), A.
[1] Sur une extension de l'intégrale de M. Denjoy. *Compt. Rend.* **162**, 287–291 (1916)
[2] Sur le procédé d'integration de M. Denjoy. *Rec. Math. Soc. Math., Moscow,* **30**, 548–557 (1918).

LEBESGUE, H.
[1] Sur une généralisation de l'intégrale définie. *Compt. Rend.* **132**, 1025–1028 (1901).
[2] Intégrale, Longueur, Aire. *Ann. Mat. Puza Appl.* **7**, (3) 231–359 (1902). (Also published separately in Paris.)
[3] "Leçons sur l'Intégration et la Recherche des Fonctions Primitives, Paris, 1904.
[4] "Leçons sur l'Integration et la Recherche des Fonctions Primitives," 2nd ed. Gauther-Villars, Paris, 1928.
[5] Russian transl. of [4]. Moscow, 1934.
[6] Sur les fonctions représentables analytiquement. *J. Math.* **1** (6), 139–216 (1905).
[7] Sur une propriété des fonctions, *Compt. Rend.* **137**, 966–968 (1903).
[8] "Leçons, sur les Séries Trigonométriques." Paris, 1905.
[9] Sur les fonctions dérivées. *Atti Acad. Naz. Lincei Rend.* **15**, 3–8 (1906).
[10] Sur la recherche des fonctions primitives par l'intégration. *Atti Acad. Naz. Lincei Rend.* **16**, 283–290 (1907).

[11] Sur l'intégrale de Stielties et sur les équations fonctionelles linéaires. *Compt. Rend.* **150**, 86–88 (1910).

[12] Sur l'intégration des fonctions discontinués. *Ann. Ecole Norm. Sup.* **27**, (3) 361–450 (1910).

[13] Remarques sur les théories de la mesure et de l'intégration. *Ann. Ecole Norm. Sup.* **35** (3), 191–250 (1918).

[14] Sur une définition due à M. Borel. *Ann. Ecole Norm. Sup.* **37**, (3) 255–257 (1920).

[15] Sur les intégrales singulières. *Ann. Fac. Sci. Toulouse* **1** (3), 4–117 (1909).

LEJEUNE–DIRICHLET, G.

[1] Sur la convergence des séries trigonométriques qui servent à représenter une fonction arbitraire entre des limites données. *J. Math.* **4**, 157–169 (1829).

[2] Sur les séries dont le terme général dépend de deux angles, et qui servent à exprimer des fonctions arbitraires entre des limites données. *J. Math.* **17**, 35–56 (1837).

LEVI, B.

[1] Richerche sulle funzioni derivate. *Atti. Accad. Naz. Lincei, Rend.* **15**, 433–438, 551–558, 674–684 (1906).

LIPSCHITZ, P.

[1] De explicatione per series trigonometricas instituenda functiorum unius variabilis arbitrariarum...., *J. Math.* **63**, 296–308 (1864).

[2] De explicatione per series trigonometricas instituenda functiorum unius variabilis arbitrariarum . . . (French transl. by P. Montel). *Acta Math.* **36**, 282–295 (1912–1913).

LOOMAN, H.

[1] Über die Perronsche Integraldefinition. *Math. Ann.* **93**, 153–156 (1925).

LUSIN, N.

[1] On the basic theorem of the integral calculus, *Matem. Sbornik* (*Rec. Math.*) **28**, 266–294 (1912). (In Russian.)

[2] Sur les propriétés des fonctions mesurables. *Compt. Rend.* **154**, 1688–1690 (1912).

[3] Sur les propriétés de l'integrale de M. Denjoy. *Compt. Rend.* **155**, 1475–1477 (1912).

[4] The integral and trigonometric series. Thesis, Moscow, 1915. (in Russian.)

[5] The present state of the theory of functions of real variables. *Trans. Russ. Math. Congr. Moscow*, 1927. (in Russian.)

[6] "The integral and Trigonometric Series." Moscow–Leningrad, 1951.

MEN'SHOV, D. E.

[1] Sur le rapport entre les définitions de l'intégrale de M. Borel et de M. Denjoy. *Rec. Math.* **30** (1916).

MOORE, E. N.
[1] Concerning Harnack's theory of improper definite Integrals. *Trans. Amer. Math. Soc.* **2**, 296–330 (1901).

NATANSON, I. P.
[1] "Theory of Functions of a Real Variable," Chapters I–IX, 1955, Chapters X–XVII; 1959; Ungar, New York, 1st Russ. ed., 1950, 2nd Russ. ed., 1957, Moscow–Leningrad.

PEANO, G.
[1] "Applicationi Geometriche del Calcolo Infinitesimale." Torino, 1887.

PERRON, O.
[1] Über den Integralbegriff. *Sitzber. Heidelberg. Akad. Wiss. Abt.* **A16**, 1–16 (1914).

PIERPONT, J.
[1] "Lectures on the Theory of Functions of Real Variables," Vol. I. Ginn, Boston, 1950.
[2] "Lectures on the Theory of Functions of Real Variables," Vol. II.
[3] Reply to Prof. Frechet's article. *Bull. Amer. Math. Soc.* **22**, 298–302 (1916).

POLLARD, S.
[1] The Stieltjes integral and its generalizations, *Quart. J. Math. Oxford Ser.* **49**, 87–94 (1920).

RADON, J.
[1] Theorie und Anwendungen der absolut additiven Mengenfunktionen. *Sitzber. Akad. Wiss. Wien* **122**, IIa, 1295–1438 (1913).

RIEMANN, B.
[1] Ueber die Darstellbarkeit einer Funktion durch eine trigonometrische Reihe. *Abh. Kön. Ges. Wiss. Göttingen* **13** (1867).
[2] "Gesammelte Mathematische Werke" (Collected works of Bernhard Riemann) (H. Weber, ed.), 1892 (Suppl., 1902); 2nd ed., Dover, 1953; Russ. ed., Moscow, 1948.

RIESZ, F.
[1] Sur quelques points de la théorie des fonctions sommables. *Compt. Rend.* **154**, 641–643 (1912).
[2] Sur les opérations fonctionnelles linéaires. *Compt. Rend.* **149**, 974–977 (1909).

RIESZ, F., and SZ.-NAGY, B.
[1] "Functional Analysis." Ungar, New York, 1955.

SAKS, S.
[1] "Theory of the Integral." Warsaw, 1937. Hafner Publ. Co.

SCHOENFLIES, A.
[1] Die Entwicklung der Lehre von den Punktmannigfaltigkeiten. *Jahresber. Deutsch. Math. Ver.* **8** (1900).
[2] "Entwicklung der Mangenlehre und ihrer Anwendungen, erste Hälfte": Allgemeine Theorie der unendlichen Mengen und Theorie der Punktmengen. Leipzig–Berlin, 1913.

SMITH, H. J., SR.
[1] On the integration of discontinuous functions. *Proc. Lond. Math. Soc.* **6** (1), 148–160 (1875).

STIELTJES, T. J.
[1] Recherches sur les fractions continues. *Ann. Fac. Sci. Toulouse* **8** (1), 1–122 (1895).

STOLZ, O.
[1] Über einen zu einer unendlichen Punktmenge gehörigen Grenzwert. *Math. Ann.* **23** 152–156 (1884).
[2] "Grundzüge der Differenzial- und Integralrechnung," Vol. III. Leipzig, 1899.

VITALI, G.
[1] Una proprietà della funzioni misurabili. *Reale Ist. Lombardo, Rend.* **38**, (2), 600–603 (1905).
[2] Sulle funzioni integrali. *Atti. Accad. Sci. Torino* **40**, 1021–1034 (1904–1905).
[3] Sui gruppi di punti. *Rend. Circ. Math. Palermo* **18**, 116–126 (1904).

YOUNG, W. H.
[1] On the general theory of integration. *Phil. Trans. Roy. Soc. London* **204A**, 221–252 (1905).
[2] Open sets and the theory of content. *Proc. London Math. Soc.* **2** (2), 16–51 (1904).
[3] On functions and their associated sets of points. *Proc. London Math. Soc.* **12** (2), 260–287 (1912–1913).
[4] On a new method of integration. *Proc. London Math. Soc.* **9**, (2) 15–50 (1910).
[5] On the new theory of integration. *Proc. Roy. Soc. London* **88A**, 170–178 (1912).
[6] Integration with respect to a function of bounded variation. *Proc. London Math. Soc.* **13** (2), 109–150 (1914).
[7] On non-absolutely convergent, not necessarily continuous integrals. *Proc. London Math. Soc.* **16** (2), 175–218 (1917–1918).

YUSHKEVICH, A. P.
[1] On the origin of the notion of Cauchy's definite integral. *Trans. Inst. History Natural Sci. USSR* **1** (1947) (in Russian).

ZAANEN, A. C.
[1] "An Introduction to the Theory of Integration." North-Holland Publ., Amsterdam, 1958 (2nd rev. ed., 1968).

AUTHOR INDEX

This index—prepared especially for the English edition—is intended as an auxiliary to the detailed Table of Contents and Bibliography. It contains references only to those occurrences not easily identifiable from the Table of Contents. For example, there are no references to W. H. Young for any pages in Chapter 5, which is devoted to Young's integral.

For the sake of historical comparisons, the dates of birth (and death where appropriate) supplement the listings of most of the authors. (*Translator's remarks.*)

SUBJECT INDEX

This index—prepared especially for the English edition—is intended as an auxiliary to the detailed Table of Contents. It contains references only to those occurrences not easily identifiable from the Table of Contents.

For example, there is no reference to Young's integral for any pages in Chapter 5, which is wholly devoted to this topic. The terms defined in the introductory section (Notation and Terminology) are not included in this list. (*Translator's remarks.*)